John Cameron

Gaelic Names of Plants (Scottish and Irish)

John Cameron

Gaelic Names of Plants (Scottish and Irish)

ISBN/EAN: 9783744730112

Printed in Europe, USA, Canada, Australia, Japan

Cover: Foto ©berggeist007 / pixelio.de

More available books at **www.hansebooks.com**

GAELIC NAMES OF PLANTS

" I study to bring forth some acceptable work : not striving to shew any rare invention that passeth a man's capacity, but to utter and receive matter of some moment known and talked of long ago, yet over long hath been buried, and, as it seemed, lain dead, for any fruit it hath shewed in the memory of man."— *Churchward*, 1588.

GAELIC NAMES OF PLANTS

(SCOTTISH AND IRISH)

COLLECTED AND ARRANGED IN SCIENTIFIC ORDER, WITH
NOTES ON THEIR ETYMOLOGY, THEIR USES, PLANT
SUPERSTITIONS, ETC., AMONG THE CELTS,
WITH COPIOUS GAELIC, ENGLISH,
AND SCIENTIFIC INDICES

BY

JOHN CAMERON
SUNDERLAND

"WHAT'S IN A NAME? THAT WHICH WE CALL A ROSE
BY ANY OTHER NAME WOULD SMELL AS SWEET."
—*Shakespeare.*

WILLIAM BLACKWOOD AND SONS
EDINBURGH AND LONDON
MDCCCLXXXIII

PREFACE.

THE Gaelic Names of Plants, reprinted from a series of articles in the 'Scottish Naturalist,' which have appeared during the last four years, are published at the request of many who wish to have them in a more convenient form. There might, perhaps, be grounds for hesitation in obtruding on the public a work of this description, which can only be of use to comparatively few; but the fact that no book exists containing a complete catalogue of Gaelic names of plants is at least some excuse for their publication in this separate form. Moreover, it seemed to many able botanists that, both for scientific and philological reasons, it would be very desirable that an attempt should be made to collect such names as are still used in the spoken Gaelic of Scotland and Ireland, before it became too late by the gradual disappearance of the language. Accordingly the author undertook this task at the request of the Editor of the 'Scottish Naturalist,' Dr Buchanan White, F.L.S. If the difficulties of its accomplishment had been foreseen, he would have hesitated to make the attempt ; as it is, nearly ten years of his life have been occupied in searching through vocabularies, reading Irish and Scottish Gaelic, and generally trying to bring into order the confusion to which these names have been reduced partly by the carelessness of the compilers of Dictionaries, and frequently by their botanical ignorance. To accomplish this, numerous journeys had to be undertaken among the Gaelic-speaking populations, in order, if possible, to settle disputed names, to fix the plant to which the name was applied, and to collect others previously unrecorded.

In studying the Gaelic nomenclature of plants, it soon became evident that no collection would be of any value unless the Irish-Gaelic names were incorporated. Indeed, when the lists supplied by Alexander M'Donald (*Mac-Mhaighster-Alastair*), published in his vocabulary in 1741, are examined, they are found to correspond with those in much older vocabularies published in Ireland. The same remark applies, with a few exceptions, to the names of plants in Gaelic supplied by the Rev. Mr Stewart of Killin, given in Lightfoot's 'Flora Scotica.' Undoubtedly, the older names have been preserved in the more copious Celtic literature of Ireland; it is certainly true that "*In vetustâ Hibernicâ fundamentum habet.*" The investigations of Professor O'Curry, O'Donovan, and others, have thrown much light on this as well as upon many other Celtic topics. The Irish names are therefore included, and spelt according to the various methods adopted by the different authorities; this gives the appearance of a want of uniformity to the spelling not altogether agreeable to Gaelic scholars, but which, under the circumstances, was unavoidable.

It was absolutely essential that the existing Gaelic names should be assigned correctly. The difficulty of the ordinary botanical student was here reversed: he has the plant but cannot tell the name — here the name existed, but the plant required to be found to which the name applied. Again, names had been altered from their original form by transcription and pronunciation; it became a matter of difficulty to determine the *root* word. However, the recent progress of philology, the knowledge of the laws that govern the modifications of words in the brotherhood of European languages, when applied to these names, rendered the explanation given not altogether improbable. Celts named plants often from (1), their uses; (2), their appearance; (3), their habitats; (4), their superstitious associations, &c. The knowledge of this habit of naming was the key that opened many a difficulty.

For the sake of comparison a number of Welsh names is given, selected from the oldest list of names obtainable, —those appended to Gerard's 'Herbalist,' 1597.

The author cannot sufficiently express his obligation to numerous correspondents in the Highlands and in Ireland for assistance in gathering local names; without such help it would have been impossible to make a complete collection. Notably the Rev. A. Stewart, Nether Lochaber, whose knowledge of natural history is unsurpassed in his own sphere ; the Very Rev. Canon Bourke, Claremorris, who gave most valuable assistance in the Irish names, particularly in the etymology of many abstruse terms, his accurate scholarship, Celtic and classical, helping him over many a difficulty; Mr W. Brockie, an excellent botanist and philologist, who some years ago made a collection of Gaelic names of plants which was unfortunately destroyed, placed at the author's disposal valuable notes and information relative to this subject ; and lastly, the accomplished Editor of the 'Scottish Naturalist,' who, from its commencement, edited the sheets and secured the correct scientific order of the whole.

With every desire to make this work as free from errors as possible, yet, doubtless, some have escaped attention ; therefore, any names omitted, any mistake in the naming of the plants, or any other fact tending towards the further elucidation of this subject will be thankfully received for future addition, correction, or amendment.

JOHN CAMERON.

SUNDERLAND, *January* 1883.

THE GAELIC NAMES OF PLANTS.

Thalictrum—(θαλλω, *thallo*, to grow green).

Gaelic : *rugh, rù, ruigh,* ⎫
Irish : *ruib,*   ⎬ Rue (or plants resembling *Ruta*
graveolens). See Gerard. ⎭

T. alpinum.—*Rú ailpeach :* Alpine meadow-rue.

T. minus.—*Rú beg :* Lesser meadow-rue. RUE is nearly the
same in most of the ancient languages; said to be from ρυω,
to flow ; Gaelic — *ruith*, flow, rush ; their roots, especially
T. flavum, possessing powerful cathartic qualities like *rhubarb*.
Compare also *ru, run*, a secret, mystery, love, desire, grace.
Welsh : *runa*, hieroglyphics (Runic). The Thalictrum of Pliny
is supposed to be the *meadow-rue*. (See Freund's Lexicon.)

> "I'll set a bank of rue, sour herb of grace"—SHAKESPEARE.

> "Mo *rùn* geal og !"—My fair young *beloved* one !

> "Oir a ta sibh a toirt deachaimh 'a mionnt, agus a *rù*, agùs gach uile ghnè
> luibhean."—For ye tithe mint and *rue*, and all manner of herbs.

The Rue of Shakespeare is generally supposed to be *Ruta
graveolens* (*Rù gharaidh*), a plant belonging to another order,
and not indigenous.

Anemone nemorosa—Wind-flower. Gaelic : *plùr na gaoithe*,
wind-flower (Armstrong). Welsh : *llysiau'r gwynt*, wind-flower,
because some of the species prefer windy habitats. Irish :
nead chailleach, old woman's nest.

Ranunculus.—From Gaelic, *ran ;* Egyptian, *ranah ;* Latin,
rana, a frog, because some of the species inhabit humid places
frequented by that animal, or because some of the plants have
leaves resembling in shape a frog's foot. *Ranunculus* is also

sometimes called crowfoot. Gaelic : *cearban*, raggy, from its divided leaves. *Gair-cean*,—from *gair*, a smile ; *cean*, love, elegance. Welsh : *crafrange y frân*, crows' claws.

R. aquatilis—Water crowfoot. Gaelic: *fleann uisge*, probably from *lean*, to follow, and *uisge*, water, follower of the water. *Lìon na h'aibhne*, the river-flax. Irish : *neul uisge*,—*neul*, a star, and *uisge*, water. *Tuir chis*,—*tuir*, a lord ; *chis*, purse (from its numerous achenes).

R. ficaria—Lesser celandine. Gaelic : *grain-aigein*, that which produces loathing. *Searraiche*, a little bottle, from the form of the roots. Welsh : *toddedig wen*, fire dissolvent ; *toddi*, melt, dissolve.

R. flammula — Spearwort. Gaelic : *glas-leun*,—*glas*, green ; *leun*, a swamp. *Lasair-leana*,—*lasair*, a flame, and *leana* or *leun*, a swamp, a spear. Welsh : *blaer y guaew*, lance-point.

R. auricomus — Goldilocks. Gaelic : *follasgain* ; probably from *follais*, conspicuous. Irish : *foloscain*, a tadpole. The Gaelic may be a corruption from the Irish, or *vice versa* ; also *gruag Mhuire*, Mary's locks.

R. repens—Creeping crowfoot. Gaelic : *buigheag*, the yellow one. Irish : *bairgin*, more frequently *bairghin*, a pilgrim's habit. *Fearban*,—*fearba*, killing, destroying.

R. acris—Upright meadow crowfoot. Gaelic : *cearban feoir*, the grass rag. Irish : the same name. This plant and *R. flammula* are used in the Highlands, applied in rags (*cearban*), for raising blisters.

R. bulbosus—Bulbous crowfoot. Gaelic : *fuile thalmhainn*, blood of the earth (it exhausts the soil). Welsh : *crafange y frân*, crows' claws.

R. sceleratus — Celery-leaved crowfoot. Gaelic and Irish : *torachas biadhain ;* probably means food of which one would be afraid.

Caltha palustris—Marsh-marigold. Gaelic : *a chorrach shod*, the clumsy one of the marsh. *Lus bhuidhe bealtuinn*, the yellow plant of Beltane or May,—*Bel* or *Baal*, the sun-god, and *teine*, fire. The name survives in many Gaelic names—*e.g.*, *Tullibeltane*, the high place of the fire of *Baal*.

" Beath a's calltuinn latha-*bealthuinn*."—M'Kay.
Birch and hazel first day of May.

Irish : *plubairsin* from *plubrach*, plunging. *Lus Mairi*, Marywort, Marygold.

Helleborus viridis — Green hellebore. Gaelic : *elebor*, a corruption of *helleborus* (from the Greek ἑλειν, *helein*, to cause death ; and βορα, *bora*, food—poisonous food).

> "Mo shròn tha stocpt à dh'*elebor*."—M'DONALD.
> My nose is stopped with hellebore.

H. fœtidus — Stinking hellebore. *Meacan sleibhe*, the hill-plant.

Aquilegia vulgaris—Columbine. Gaelic : *lus a cholamain*, the dove's plant. Irish : *cruba-leisin*,—from *cruba*, crouching, and *leise*, thigh or haunch; suggested by the form of the flower. *Lusan cholam* (O'Reilly), pigeon's flower. Welsh : *troed y glomen*, naked woman's foot.

Aconitum napellus—Monkshood. Gaelic : *fuath mhadhaidh* (Shaw), the wolf's aversion. *Curaichd mhànaich* (Armstrong), monkshood. Welsh : *bleiddag*,—from *bleidd*, a wolf, and *tag*, choke.

Nigella damascena—Chase-the-devil. Gaelic : *lus an fhograidh*, the pursued plant. Irish : *lus mhic Raonail*, MacRonald's wort. Not indigenous, but common in gardens.

Pæonia officinalis—Peony. Gaelic : *lus a phione*. A corruption of *Pæon*, the physician who first used it in medicine, and cured Plato of a wound inflicted by Hercules. Welsh : *bladeu'r brenin*, the king's flower. Irish : *lus phoinc*.

BERBERIDACEÆ.

Berberis vulgaris—Barberry. Gaelic : *barbrag* (a corruption from *Phœnician* word *barar*), the brilliancy of a shell; alluding to their shining leaves. Greek βερβερι, *berberi*, a shell. *Preas nan gear dhearc*, the sour berry-bush. *Preas deilgneach*, the prickly bush. Irish : *barbrog*.

NYMPHÆACEÆ.

(From νυμφη, *nymphe*, a water-nymph, referring to their habitats.)

Nymphæa alba—White water-lily. Gaelic : *duileag bhaite bhàn*, the drowned white leaf.

> "Feur lochain is tachair,
> An cinn an *duileag bhàite*."—M'INTYRE.
> Water, grass, and algæ,
> Where the water-lily grows.

> "O *lili*, righ nam fleuran."—M'DONALD.
> O lily, king of flowers.

Rabhagach, giving caution or warning; a beacon. *Lili bhàn*, white lily. Welsh: *Lili-r-dwfr*, water-lily. Irish: *buillite*. (Shaw.) **Nuphar luteum** — Yellow water-lily. Gaelic: *duileag bhaite bhuidhe*, the yellow drowned leaf. *Lili bhuidhe n'uisge*, yellow water-lily. Irish: *liach laghor*, the bright flag. *Cabhan abhain*,— *cabhan*, a hollow plain, and *abhain*, of the river.

PAPAVERACEÆ.

Papaver rhœas — Poppy. Gaelic: *meilbheag*, sometimes *beilbheag*, a little pestle (to which the capsule has some resemblance).

"Le *meilbheag*, le noinean, 's le slan-lus."—M'LEOD.

With a poppy, daisy, and rib-grass.

Fothros, corn-rose,—from *ioth* (Irish), corn ; *ros*, rose. *Cromlus*, bent weed. *Paipean ruadh*,—*ruadh*, red, and *paipean* a corruption of *papaver*, from *papa*, pap, or *pappo*, to eat of pap. The juice was formerly put into children's food to make them sleep. Welsh : *pabi*.

P. somniferum—Common opium poppy. Gaelic : *codalian*, from *codal* or *cadal*, sleep.

Chelidonium majus. Common celandine (a corruption of χελιδων, *chelidon*, a swallow). Gaelic : *an ceann ruadh*,[1] the red head. Irish : *lacha cheann ruadh*, the red-headed duck. Welsh : *llysie y wennol*, swallow-wort. The flower is yellow, not red. *Aonsgoch* is another Gaelic name for swallow-wort, meaning the lonely flower,—*aon*, one or alone, and *sgoth*, a flower.

Glaucium luteum — Yellow horned poppy. Gaelic : *barrag ruadh* (?), the valiant or strong head. The flower is yellow, not red.

FUMARIACEÆ.

(From *fumus*, smoke. "The smoke of these plants being said by the ancient exorcists to have the power of expelling evil spirits" (Jones). French : *fume terre*.)

Fumaria officinalis — Fumitory. Gaelic : *lus deathach thalmhainn* (Armstrong), the earth-smoke plant. Irish : *deatach thalmhuin* (O'Reilly), earth-smoke. Welsh : *mwg y ddaer*, earth-smoke. Another Irish name is *caman scarraigh* (O'Reilly), —*caman*, crooked, and *scaradh*, to scatter.

---

[1] *Ruadh* does not mean absolutely red, but reddish. Welsh : *Rhydh*. It means also power, virtue, strong, valiant.

CRUCIFERÆ.

(From Latin *crux, crucis*, a cross, and *fero*, to bear, the petals being arranged crosswise.)

Crambe maritima—Seakale. Gaelic: *praiseag tràgha*, the shore pot-herb,—from the Irish *praiseach*, Gaelic *praiseag*, a little pot (a common name for pot-herbs). *Càl na màra*, sea-kale (from Greek, χαυλος; Latin, *caulis;* German, *kohl;* Saxon, *cawl;* English, *cole* or *kale;* Irish, *càl;* Welsh, *cawl.*)

Isatis tinctoria — Woad. The ancient Celts used to stain their bodies with a preparation from this plant. Its pale-blue hue was supposed to enhance their beauty, according to the fashion of the time. Gaelic: *guirmean*, the blue one. Irish and Gaelic: *glas lus*, pale-blue weed. Welsh: *glas lys*. Formerly called *Glastum.*

" Is *glas* mo luibh."—OSSIAN.
Pale-blue is the subject of my praise.

On account of the brightness of its manufactured colours the Celts called it *gwed* (*guède* in French to this day), whence the Saxon *wad* and the English *woad*.

Thlaspi arvense—Penny-cress. Gaelic: *praiseach fcidh*, deer's pot-herb. Irish: *preaseach fiadh*, a deer's pot-herb.

Capsella Bursa-pastoris—Shepherd's purse. Gaelic: *lus na fola*, the blood-weed; *an sporran*, the purse. Irish: *sraidin*, a lane, a walk. Welsh: *purs y bugail*, shepherd's purse (*bugail*, from Greek βυκολος, a shepherd).

Cochlearia officinalis — Scurvy-grass. Gaelic: *a maraich*, sailor; *carran*, the thing for scurvy, possessing antiscorbutic properties. "*Plaigh na carra*," the plague of leprosy (Stuart). "*Duine aig am bheil carr*," a man who has the scurvy (Stuart in Lev.) Welsh: *mor luyau*, sea-spoons; *llysie'r blwg*, scurvy-grass (from *blwg*, scurvy). Irish: *biolair tràgha,—biolair*, dainty, and *tràgha*, shore or seaside.

Armoracia rusticana (*armoracia*, a name of Celtic origin,—from *ar*, land; *mor* or *mar*, the sea; *ris*, near to,—a plant growing near the sea). English: *horse-radish*. Gaelic: *meacan-each*, the horse-plant. Irish: *racadal*, perhaps from an old word *rac*, a king, a prince, and *adhal*, desire—*i.e.*, the king's desire.

Raphanus raphanistrum—Radish. Gaelic: *meacan ruadh*, the reddish plant, from the colour of the root. Irish: *fiadh roidis*, wild radish.

Cardamine pratensis—Cuckoo flower, ladies' smock. Gaelic: *plur na cubhaig*, the cuckoo-flower. *Gleoran*, from *gleote*, handsome, pretty. The name is given to other cresses as well. *Biolair-ghriagain*, the bright sunny dainty.

Cakile maritimum—Sea gilly-flower rocket. Gaelic: *fearsaid-eag;* meaning uncertain, but probably from Irish *saide*, a seat (Latin, *sedes*), the sitting individual—from its procumbent habit.

Nasturtium officinalis—Water-cress. Gaelic, *biolair*, a dainty, or that which causes the nose to smart, hence agreeing with *nasturtium* (Latin : *nasus*, the nose, and *tortus*, tormented). *Durlus,—dur*, water, and *lus*, plant. *Dobhar-lus,—dobhar*, water. Welsh : *berwyr dwfr*, water-cress. The Gaelic and Irish bards used these names indefinitely for all cresses.

> " Sa *bhiolair* luidneach, shliòm-chluasach.
> Glas, chruinn-cheannach, chaoin ghorm-nealach';
> Is i fàs glan, uchd-ard, gilmeineach,
> Fuidh barr geal iomlan, sonraichte."—M'INTYRE.

Its drooping, smooth, green, round-leaved water-cress growing so radiantly, breast-high, trimly; under its remarkably perfect white flower.

> "*Dobhrach* bhallach mhìn."—M'INTYRE.

Smooth-spotted water-cress.

Sisymbrium sophia—Flixweed. Gaelic : *fineal Mhuire*, the Virgin Mary's fennel. Welsh : *piblys*, pipe-weed.

Erysimum alliaria—Garlic mustard, sauce alone. Gaelic : *garbhraitheach*, rough, threatening.

Cheiranthus cheiri—Wallflower, gilly-flower. Gaelic : *lus leth an samhraidh*, half the summer plant. Irish : the same. Welsh : *bloden gorphenaf*, July flower or gilly-flower. Wedgwood says gilly-flower is from the French *giroflée*.

Brassica rapa—Common turnip. Gaelic, *neup;* Irish, *neip ;* Welsh, *maipen ;* Scotch, *neep* (and *navew*, French, *navet*); corruptions from Latin *napus.*

B. campestris—Wild navew. Gaelic : *neup fiadhain*, wild turnip.

B. oleracea—Seakale or cabbage. Gaelic and Irish : *praiseach bhaidhe*, the pot-herb of the wave (*baidhe*, in Irish, a wave. *Morran,—mor* (Welsh), the sea, its habitat the seaside. *Cal colbhairt*—the kale with stout fleshy stalks (from *colbh*, a stalk of a plant, and *art*, flesh), *càl* or *cadhal*. Welsh : *cawl*, kale. Gaelic : *càl-cearslach* (*cearslach*, globular), cabbage ; *càl gruidhean* (with grain like flowers), cauliflower ; *colag* (a little cabbage), cauliflower ; *garadh càil*, a kitchen-garden.

> " Dh' itheadh biolair an fhuarain
> 'S air bu shuarach an *càl*."—M'DONALD.
> I would eat the cress of the wells.
> Compared to it, kale is contemptible.

Sinapis arvensis—Charlock, wild mustard. Gaelic : *marag bhuidhe*, the yellow sausage (to which the pod is supposed to bear some resemblance). *Sceallan,—sceall*, a shield. *Sgealag* (Shaw),—*sgealpach*, biting. *Mustard*, from the English.

> " Mar ghrainne de shìol *mustaird*."—STUART.
> Like a grain of mustard-seed.

Gaelic : *praiseach garbh*, the rough pot-herb.

RESEDACEÆ.

Reseda luteola—Weld, yellow weed. Gaelic : *lus buidhe mòr*, the large yellow weed. Irish : *buidhe mòr*, the large yellow. Welsh : *llysie lliu*, dye-wort. *Reseda*, from Latin *resedo*. Gaelic : *reidh*, to calm, to appease.

CISTACEÆ.

(From Greek κιστη, *kiste*, a box or capsule, from their peculiar capsules. Latin, *cista ;* Gaelic, *ciste ;* Danish, *kiste*.)

Helianthemum vulgare—Rock-rose. Gaelic : *grian ròs*, sun-rose ; *plùr na gréine*, flower of the sun (also heliotrope). Welsh : *blodaw'r haul*, sun-flower.

VIOLACEÆ.

(From Greek ἴον, *ion*, a violet,—the food given to the cow Io, one of Jupiter's mistresses.)

Viola odorata—Sweet violet. Gaelic : *fail chuach*, scented bowl ; *fail*, scent, and *cuach*, a bowl hollow as a nest. Scotch : *quaich*, *cogie* (dim.), a drinking-cup.

> " *Fail chuachaig* ar uachdar a fheoir."—M'FARLANE.
> Scented violet on the top of the grass.

V. canina—Dog-violet. Gaelic : *dail chuach*, field-bowl (*dail*, a field). Danish : *dal*, a valley.

> " Gun sobhrach gun *dail chuach*,
> Gun lus uasal air càrnn."—M'INTYRE.
> Without primrose or violet,
> Or a gay flower on the heap.

Sàil chuach,—sail, a heel (from its spur).

"Coille is guirme *sàil chuach.*"—Old Song.
A wood where violets are bluest.

Irish : *biodh a leithid,* the world's paragon ; also *fanaisge,* probably from *fan,* weak, faint, agreeing in meaning with the Welsh name, *crinllyns,* a fragile weed.

DROSERACEÆ.

(From Greek δροσερος, *droseros,* dewy, because the plants appear as if covered with dew.)

Drosera rotundifolia — Round-leaved sundew. Gaelic : *ròs an t'solais,* sun-rose or flower ; *geald-ruidhe* or *dealt ruaidhe,* very red dew ; *lus na fearnaich,* the plant with shields (its leaves have some resemblance to shields). Irish : *eil druich* (*eil,* to rob, and *druich,* dew), the one that robs the dew ; *druichdin mona,* the dew of the hill. Welsh : *doddedig rudd,*—*dod,* twisted thread, and *rudd,* red, the plant being covered with red hairs.

POLYGALACEÆ.

(From Greek πολυ, *poly,* much, and γαλα, *gala,* milk.)

Polygala vulgaris—Milkwort. Gaelic : *lus a bhàine,* milkwort. Irish : *lusan baine,* the same meaning, alluding to the reputed effects of the plants on cows that feed upon it.

CARYOPHYLLACEÆ.

Saponaria officinalis—Soapwort, bruisewort. Gaelic: *gairgeancregach.* Irish : *gairbhin creugach,* the bitter one of the rocks ; *garbhion,* bitterness, and *creugach,* rocky. The whole plant is bitter, and was formerly used to cure cutaneous diseases. *Lus an shiabunn,* the soapwort. Welsh : *sebonllys,* the same meaning (*sebon,* soap), Latin *sapo,* so called probably because the bruised leaves produce lather like soap. Soap was a Celtic invention.

"Prodest et *sapo.* Gallorum hoc inventum,
Rutilandis capillis, ex sevo et cinere."—PLINY.

Lychnis flos-cuculi—Ragged robin. Gaelic : *plur na cubhaig,* the cuckoo flower ; *curachd na cubhaig,* the cuckoo's hood.

L. diurna—Red campion. Gaelic: *àrean coileach,* cockscomb ; in some places *corcan coille,* red woodland flower.

L. githago—Corn-cockle. Gaelic: *brog na cubhaig,* the cuckoo's shoe. *Luibh laoibheach,—laoi,* day, and *beachd,* to observe—*i e.,* the plant observed for a day. Irish: *cogall,*[1] from *coch* (Welsh), red; hence *cockle.* French: *coquille.* Welsh: *gith,* cockle or its seed, a corruption from *githago,* or *vice versâ.*

Spergula arvensis—Spurrey. Gaelic: *cluain lin,—cluain,* fraud, and *lin,* flax—*i.e.,* fraudulent flax. *Carran,* twisted or knotted. Scotch: *yarr.* Irish: *cabrois,—cab,* a head; *rois,* polished.

> " Gun deanntag, gun charran."—M'DONALD.
> Without nettle or spurrey.

Arenaria alsine—Sandwort. Gaelic: *flige,* perhaps from *fliche,* water, growing in watery or sandy places.

Stellaria media—Chickweed. Gaelic: *fliodh,* an excrescence (Armstrong), sometimes written *fluth.* Irish: *lia,* wetting (Gaelic: *fluich,* wet); compare also *floch,* soft (Latin: *flaccus*). Welsh: *gwlydd,* the soft or tender plant.

S. Holostea—The greater stitchwort. Gaelic: *tuirseach,* sad, dejected. Irish: *tursarrain,* the same meaning; and **Stellaria graminea,** *tursarranin,* the lesser stitchwort. Welsh: *y wenn-wlydd,* the fair soft-stemmed plant, from *gwenn* and *gwlydd,* soft tender stem.

Cherleria sedoides—Mossy cyphel, found plentifully on Ben Lawers. No Gaelic name, but *seorsa còinich,* a kind of moss.

Cerastium alpinum—Mouse-ear chickweed. Gaelic: *cluas an luch,* mouse-ear.

LINACEÆ.

Linum usitatissimum—Flax. Gaelic: *lìon,* gen. singular *lìn.* Welsh: *llin.* " Greek λίνον and Latin *linum,* a thread, are derived from the Celtic."—LOUDON.

> " Iarraidh i olan agus lìon."—STUART (Job).
> She will desire wool and flax.

L. catharticum—Fairy flax. Gaelic: *lìon na bean sìth,* fairy woman's flax; *miosach,* monthly, from a medicinal virtue it was supposed to possess; *mionach,* bowels; *lus caolach,* slender weed: compare also *caolan,* intestine (Latin: *colon,* the large intestine). Both names probably allude to its cathartic effects. Stuart, in Lightfoot's 'Flora,' gives these names in a combined form,—*an caol miosachan,* the slender monthly one. Irish: *ceo-lagh.*

---

[1] This plant is sometimes called *Curach na Cubhaig,* and *Cochal*—(hood or cowl). Latin: cucullus.

MALVACEÆ.

Latin : *malvæ*, mallows. Gaelic : *maloimh*, from Greek μαλάχη, *malache*, soft, in allusion to the soft mucilaginous properties of the plants.

"A' gearradh sios *maloimh* laimh ris na preasaibh, agus freumhan aiteil mar bhiadh."—STUART (Job xxx. 4).

"Who cut up *mallows* by the bushes, and juniper roots for their meat."

Welsh : *meddalai*, what softens. Gaelic : *mil mheacan*, honey-plant ; *gropais* or *grobais* (M'Donald) from Gothic, *grob*, English, *grub*, to dig. The roots were dug, and boiled to obtain mucilage.

Malva rotundifolia—Dwarf mallow. Gaelic and Irish : *ucas frangach*,—*ucas* from Irish *uc*, need, whence *uchd*, a breast (Greek, ὄχθη)—the mucilage being used as an emollient for breasts—and *frangach*, French—*i.e.*, the French mallow.

M. sylvestris—Common mallow. Gaelic : *ucas fheadhair*, wild mallow.

Althæa officinalis—Marsh-mallow. Gaelic and Irish : *leamhad*, perhaps from *leamhach*, insipid ; *fochas*, itch, a remedy for the itch (*ochas*, itch). Welsh : *morhocys*,—*mor*, the sea, and *hocys*, phlegm-producer, it being used for various pulmonary complaints.

TILIACEÆ.

Tilia europea—Lime-tree, linden. Gaelic : *craobh theile*. Irish : *crann teile*,—*teile*, a corruption from *tilia*. Welsh : *pis gwydden*.

HYPERICACEÆ.

Hypericum perforatum — The perforated St John's wort. Gaelic and Irish : *eala bhuidhe* (sometimes written *eala bhi*), probably from *eal* (for *neul*), aspect, appearance, and *bhuidhe* or *bhi*, yellow.

"An *eala bhuidhe* s'an noinean bàn
S'an t'sobhrach an gleann fàs, nan luibh
Anns am faigheadh an leighe liath
Furtach fiach, do chreuch a's leòn."—COLLATH.

In the glen where the *St John's wort*, the white daisy, and the primrose grow, the grey doctor will find a valuable remedy for every disease and wound.

"The belief was common among the Caledonians that for all the diseases to which mankind is liable there grows an herb somewhere, and not far from the locality where the particular disease prevails, the proper application of which would cure it."—M'KENZIE.

" Sobhrach a's *eala bhi* 's barra neoinean."—M'INTYRE.
Primrose, *St John's wort*, and daisies.

Allas Mhuire (*Mhuire*, the Virgin Mary; *allas*, perhaps another
form of the preceding names)—Mary's image, which would agree
with the word *hypericum*. According to Linnaeus it is derived
from Greek ὑπέρ, *uper*, over, and εἰκών, *eikon*, an image—that is
to say, the superior part of the flower represents an image.

Caod aslachan Cholum chille, from *Colum* and *cill* (church, cell),
St Columba's flower, the saint of Iona, who reverenced it and
carried it in his arms (*caod*,—(Irish) *caodam*, to come, and *aslachan*,
arms), it being dedicated to his favourite evangelist St John.[1] "Formerly it was carried about by the people of Scotland as a charm
against witchcraft and enchantment" (Don). Welsh : *y fendigaid*,
the blessed plant. French : *la toute-saìne*. English : *tutsan*.
The badge of Clan M'Kinnon.

ACERACEÆ.

("*Acer*, in Latin meaning sharp, from *ac*, a point, in Celtic."—
DU THEIS.)

Acer campestris—Common maple. Gaelic and Irish : *craobh
mhalip* or *malpais* ; origin of name uncertain, but very likely
from *mal*, a satchel or a husk, from the form of its samara. Some
think the name is only a corruption of *maple*—Anglo-Saxon,
mapal. Welsh : *masarnen*. Gothic : *masloenn* (from *mas*, fat),
from its abundance of saccharine juice.

A. pseudo-platanus—Sycamore. Gaelic and Irish : *craobh sice*,
a corruption from Greek *sycaminos*. The old botanists errone-
ously believed it to be identical with the sycamine or mulberry-fig
of Palestine.

"Nam biodh agaidh creidimh, theiradh sibh ris a *chraobh shicamin* so,
bi air do spionadh as do fhreumhaibh."—STUART.

If ye had faith ye might say to this *sycamore tree*, Be thou plucked up by
the root.—St Luke xvii. 6.

Craobh pleantrinn, corruption of platanus or plane-tree. Irish :
crann bàn, white tree. *Fir chrann*, same meaning.
The badge of Clan Oliphant.

VINIFERÆ.

Vitis (from the Celtic *gwyd*, a tree, a shrub. Spanish : *vid*.
French : *vigne*).

[1] Similar ideas occur in other Irish names respecting this plant : *Beachnuadh Columcille, beachnuadh beinionn, beachnuadh firionn,—beach*, to em-
brace ; *nuadh*, new ; *beinionn*, a little woman ; *firionn*, a little man.

Vitis vinifera—Vine. Gaelic: *crann fiona, fionan.* Irish: *fion,* wine. Greek: ϝοιν-ον. Latin: *vin-um. Fion dearc,* a grape.

GERANIACEÆ.

(From Greek γέρανος, *geranos,* a crane. The long beak that terminates the carpel resembles the bill of a crane ; English : cranebill. Gaelic: *crob priachain* (Armstrong), the claw of any rapacious bird.) *Lŭs-gnà-ghorm.* (M'Kenzie.) Evergreen plant.

Geranium Robertianum—Herb Robert. Gaelic and Irish : *righeal cuil* (from *righe,* reproof, and *cuil,* fly, gnat, insect), the fly reprover. *Riaghal cuil,* also *rial chuil,* that which rules insects ; *Earbull righ* (*earbull,* a tail).

"Insects are said to avoid it."—DON.

Ruidel, the red-haired. *Lus an Eallan,* the cancer weed. *Righeal righ.* Irish: *righean righ,* that which reproves a king (*righ,* a king), on account of its strong disagreeable smell. Welsh : *troedrydd,* redfoot. *Llysie Robert,* herb Robert.

G. sanguineum—Bloody cranesbill. Gaelic : *creachlach dearg,* the red wound-healer (*creach,* a wound). *Geranium Robertianum* and *Geranium sanguineum* have been and are held in great repute by the Highlanders, on account of their astringent and vulnerary properties.

OXALIDACEÆ.

(From Greek ὀξύς, *oxys,* acid, from the acid taste of the leaves.)

Oxalis acetosella—Wood-sorrel. Gaelic : *samh,* shelter. It grows in sheltered spots. Also the name given to its capsules. Also summer. It may simply be the summer flower.

"Alg itheach *saimh,*" eating sorrel.

Seamrag. Irish : *seamrog* (shamrock) (*seam,* mild and gentle), little gentle one. Referring to its appearance.

" Le-*seamragan* 's le neonainean,
'S'gach lus a dh'fheudain ainmeachadh
Cuir anbharra dhreach boidhchead air."—M'INTYRE.

With wood-sorrel and with daisies,
And plants that I could name,
Giving the place a most beautiful appearance.

Surag, the sour one ; Scotch : *sourock* (from the Armoric *sur,* Teutonic *suer,* sour). Welsh : *suran y gog,* cuckoo's sorrel.

Gaelic: *biadh nan coinean,* birds' food.   Irish: *billeog nan eun,* the leaf of the birds.

> "Timcheall thulmanan diàmhair
> Ma 'm bi'm *biadh-ionain* fàs."—M'DONALD.
>
> Around sheltered hillocks
> Where the wood-sorrel grows.

Feada coillé, candle of the woods, name given to the flower; *feadh,* a candle or rush.

> "Mar sin is leasachan soilleir,
> Do dh' *fheada-coille* na'n còs."—M'DONALD.
>
> Like the flaming light
> Of the wood-sorrel of the caverns.

CELASTRACEÆ.

Euonymus europæus — Common spindle-tree. Gaelic and Irish: *oir, feoras,—oir,* the east point, east. "*A tir an oir,*" from the land of the East (*Oirip,* Europe), being rare in Scotland and Ireland, but common on the Continent. *Oir* and *feoir* also mean a border, edge, limit, it being commonly planted in hedges. Whether the name has any reference to these significations it is very difficult to determine with certainty. *Oir,* the name of the thirteenth letter, O, of the Gaelic and Irish alphabet. It is worthy of notice that all the letters were called after trees or plants :—

	Gaelic.	English.			Gaelic.	English.
A	Ailm.	Elm.	L	.	Luis.	Quicken.
B	Beite.	Birch.	M	.	Muir.	Vine.
C	Coll.	Hazel.	N	.	Nuin.	Ash.
D	Dur.	Oak.	O	.	Oir.	Spindle-tree.
E	Eagh.	Aspen.	P	.	Peith.	Pine.
F	Fearn.	Alder.	R	.	Ruis.	Elder.
G	Gath.	Ivy.	S	.	Suil.	Willow.
H	Huath.	White-thorn.	T	.	Tin.	Heath.
I	Iogh.	Yew.	U	.	Uir.	Whitethorn.

RHAMNACEÆ.

Rhamnus (from Gaelic *ramh,* Celtic *ram,* a branch, wood).

> "Talamh nan *ramh.*"—OSSIAN.
> The country of woods.

The Greeks changed the word to ῥάμνος and the Latins to *ramus.*

R. catharticus—Prickly buckthorn. Gaelic: *ramh droighionn,* prickly wood. Welsh: *rhafnwydden,—rhaf,* to spread; *wydd,* tree.

Juglans regia—The Walnut. Gaelic : *craobh-ghallchno—gall*, a foreigner, a stranger ; *cno*, a nut.

LEGUMINIFERÆ.

Gaelic : *luis feidhleagach*, pod-bearing plants. *Bar guc*, papilionaceous flowers (Armstrong). *Por-cochullach*, leguminous.

"*Bar guc* air mheuraibh nosara."—M'INTYRE.

Blossoms on sappy branches.

Sarothamnus scoparius—Broom. Gaelic : *bealaidh* or *bealuidh* (probably from *bcal*, Baal, and *uidh*, favour), the plant that Belus favoured, it being yellow-flowered (see *Caltha palustris*). Yellow was the favourite colour of the Druids (who were worshippers of Belus), and also of the bards. Ossian describes the sun "*grian bhuidhe*," the yellow sun ; M'Intyre, his Isabel, as

" Iseabel og
An òr fhuilt *bhuidh*."

Young Isabel with the golden-yellow hair.

Irish : *brum ;* and Welsh : *ysgub*. Gaelic: *sguab*, a brush made from the broom. Latin : *scoparius*. *Giolcach sleibhe (giolc*, a reed, a cane, a leafless twig ; *sleibhe*, of the hill).

The badge of the Clan Forbes.

Cytisus laburnum — Laburnum. Gaelic : *bealuidh frangach* (in Breadalbane), in some parts *sasunach*, French or English broom (Ferguson). *Frangach* is very often affixed to names of plants of foreign origin. This tree was introduced from Switzerland in 1596. *Craobh obrun*, a corruption of laburnum.

Ulex—Name from the Celtic *ec* or *ac*, a prickle (Jones).

U. europæus—Furze, whin, gorse. Gaelic and Irish : *conasg*, from Irish *conas*, war, because of its armed or prickly appearance. Welsh : *eithin*, prickles.

" Lan *conasg* is phreasalbh."—OLD SONG.

Full of furze and bushes.

Not common in the Highlands, but plentiful about Fortingall, Perthshire.

Ononis arvensis—Rest - harrow. Gaelic and Irish : *sreang bogha*, bowstring. Welsh : *tagadr*, stop the plough ; *eithin yr eir*, ground prickles. Scotch : *cammock*, from Gaelic *cam*, crooked.

Trigonella ornithopodioides—Fenugreek, Greek hay. Gaelic : *ionntag-greugach* (Armstrong), Greek nettle ; *crubh-eoin*, Birds' shoe. Welsh : *y grog-wryan*.

Trifolium repens—White or Dutch clover. Gaelic and Irish :

seamar bhàn, the fair gentle one (see *Oxalis*); written also *sameir, siomrag, seamrag, seamrog.* Wood-sorrel and clover are often confounded, but *seamar bhàn* is invariable for white clover, and for Trifolium procumbens, hop trefoil, *samhrag bhuidhe*, yellow clover.

> " Gach *saimeir* neonean 's masag."—M'DONALD.
>
> Every clover, daisy, and berry.

> " An t-*seamrag* uine 's barr-gheal gruag,
> A's buidheann chuachach neoinein."—M'LACHUINN.
>
> The green white-headed clover.
> The yellow-cupped daisy.

The badge of Clan Sinclair.

T. pratense—Red clover. Gaelic : *seamar chapuill*, the mare's clover. *Capull*, from Greek καβάλλης, a work-horse. Latin : *caballus*, a horse. *Tri-bilean*, trefoil, three - leaved. Welsh : *tairdalen*, the same meaning. *Meillonem*, honeywort, from *mêl*, honey. Gaelic : *sùgag*, Scotch *sookie*, the bloom of clover, so called because it contains honey, and children suck it.

T. minus — Small yellow clover. Gaelic : *seangan*, small, slender.

T. arvense—Hare's-foot clover. Gaelic : *cas maidhiche* (Armstrong), hare's foot.

Lotus corniculata—Bird's-foot trefoil. Gaelic : *barra mhislean,—barra*, top or flower ; *mislean*, anything that springs or grows.

> "Glacag *misleanach*."—MACFARLANE.
>
> A grassy dell.

Anthyllis vulneraria — Kidney vetch, or Lady's Fingers. Gaelic : *meoir Mhuire*, Mary's fingers ; *cas an uain*, lamb's foot.

Vicia[1] sativa—Vetch. Gaelic and Irish : *fiatghal*, nutritious (from Irish *fiadh*, now written *biadh*, food) ; *peasair fiadhain*, wild pease ; *peasair chapuill*, mares' pease. Welsh : *idbys*, edible pease. Irish : *pis feadhain*, wild pease ; *pis dubh*, black peas.

V. cracca — Tufted vetch. Gaelic : *pesair nan luch*, mice pease ; *pesair* (Latin, *pisum ;* Welsh, *pys ;* French, *pois*, pease), are all from the Celtic root *pis*, a pea.

V. sepium—Bush vetch. Gaelic : *peasair nam preas*, the bush peas.

Lathyrus pratensis — Yellow vetchling. Gaelic : *peasair bhuidhe*, yellow peas. Irish : *pis bhùidhe*, yellow peas.

---

[1] *Vicia* (from *gwig*, Celtic, whence Greek βικιον, Latin *vicia*, French *vesce*, English *vetch*).—LOUDON.

Ervum hirsutum—Hairy vetch or tare (from *erv*, Celtic—*arv*, Latin, tilled land). Gaelic: *peasair an arbhar*, corn peas. Welsh : *pysen y ceirch,—ceirch*, oats. Gaelic : *gall pheasair*, a name for lentils or vetch. *Gall*, sometimes prefixed to names of plants having lowland habitats, or strangers.

> " Lan do *ghall pheasair*."—STUART, 2 Sam.
> Full of lentils.

Faba vulgaris—Bean. Gaelic : *ponair*. Irish : *poneir*. Cornish : *ponar* (from the Hebrew בול, *pul*, a bean (Levi). Gaelic : *ponair frangach*, French beans ; *ponair airneach*, kidney beans ; *ponair chapuill*, buckbean (*Menyanthes trifoliata*).

> " Gabh thugad fòs cruithneachd agus eorna, agus *pònair*, agus *peasair*, agus meanbh-pheasair, agus *peasair fhiadhain*, agus cuir iad ann an aon soitheach, agus dean duit fèin aran duibh."—STUART, Ezekiel iv. 9.
> " Take thou also unto thee wheat, and barley, and beans, and lentiles, and millet, and fitches, and put them in one vessel, and make thee bread thereof."

Orobus tuberosus—Tuberous bitter vetch (from Greek, ὅρω, *oro*, to excite, to strengthen, and βοῦς, an ox). Gaelic and Irish : *cairmeal* (Armstrong),—*cair*, dig; *meal*, enjoy; also *mall ;* Welsh : *moel*, a knob, a tuber—*i.e.*, the tuberous root that is dug ; *corra-meille* (M'Leod and Dewar).

> " Is clann bheag a trusa leolaicheann[1]
> Buain *corr* an co's nam bruachagan."—M'INTYRE.
> Little children gathering . . .
> And digging the bitter vetch from the holes in the bank.

Corra, a crane, and *meillg*, a pod, the crane's pod or peas. Welsh : *pys y garanod*, crane's peas ; *garan*, a crane. " The Highlanders have a great esteem for the tubercles of the roots ; they dry and chew them to give a better relish to their whisky. They also affirm that they are good against most diseases of the thorax, and that by the use of them they are enabled to repel hunger and thirst for a long time. In Breadalbane and Ross-shire they sometimes bruise and steep them in water, and make an agreeable fermented liquor with them, called *cairm*. They have a sweet taste, something like the roots of liquorice, and when boiled are well flavoured and nutritive, and in times of scarcity have served as a substitute for bread " (Lightfoot).

---

[1] *Leolaicheann*, probably *Trollius europæus* (the globe flower), from *òl*, *òlachan*, drink, drinking. Children frequently use the globe flower as a drinking-cup. Scotch : *luggie gowan*. *Luggie*, a small wooden dish ; or it may be a corruption from *trol* or *trollen*, an old German word signifying round, in allusion to the form of the flower, hence Trollius.

ROSACEÆ.

(From the Celtic. Gaelic, *rós;* Welsh, *rhos;* Armoric, *rosen;* Greek, ῥοδον: Latin, *rosa.*)

Prunus spinosa—Blackthorn, sloe. Gaelic: *preas nan airneag,* the sloe bush. Irish: *airne,* a sloe.

> " Sùilean air lidh *airneag.*"—Ross.
> Eyes the colour of sloes.

Sgitheach dùbh,—the word *sgith* ordinarily means weary, but it means also (in Irish) fear; *dubh,* black, the fearful black one, but probably in this case it is a form of *sgeach,*[1] a haw (the fruit of the white thorn), the black haw. Welsh: *eirinen ddu,* the black plum; *eiryn,* a plum.

> " Crùn *sgitheach* an aite crùn righ.—M'Ellar.
> A crown of thorns instead of a royal crown.

Droighionn dùbh, the black penetrator (from *druid,* to penetrate, pierce, bore). Compare Gothic, *thruita;* Sanscrit, *trut;* Latin, *trit;* Welsh, *draen;* German, *dorn;* English, *thorn.*

> "Croin *droignich* 'on ear's o'niar."—Old Poem.
> Thorn-trees on either side.

P. damascena—Damson. Gaelic and Irish: *daimsin* (corruption).

P. insititia — Bullace. Gaelic and Irish: *bulastair.* Compare Breton, *bolos;* Welsh, *biolas,* sloes.

P. domestica—Wild plum. Gaelic: *plumbais fiadhainn,* wild plum; *plumbais seargta,* prunes. Latin: *prunum.*

P. armeniaca—Apricot. Gaelic: *apricoc.* Welsh: *bricyllen.* Regnier supposes from the Arabic *berkoch,* whence the Italian *albicocco,* and the English *apricot;* or, as Professor Martyn observes, a tree when first introduced might have been called a "præcox," or early fruit, and gardeners taking the article "a" for the first syllable of the word, might easily have corrupted it to apricots.

P. cerasus—Cherry-tree. Gaelic: *craobh shiris,* a corruption of Cerasus, a town in Pontus in Asia, from whence the tree was first brought.

> " Do bheul mar t' *siris.*"
> Thy mouth like the cherry.

Welsh: *ceiriosen.*

---

[1] *Sgeach,* also a bush.

C

18

P. padus—Bird cherry. Gaelic : *craobh fhiodhag,* from *fiodh,* wood, timber ; *fiodhach,* a shrubbery.

P. avium—Wild cherry. Gaelic : *geanais,* the gean. French : *guigne,* from a German root.

Amydalus communis—Almond. Gaelic : *amon, cno ghreugach,* Greek nut.

A. persica—Peach. Gaelic : *peitseag,* from the English.

Spiræa ulmaria — Meadow - sweet, queen of the meadow. Gaelic : *crios* (or *cneas*) *Chu-chulainn.*[1] The plant called " My lady's belt " (M'Kenzie). " A flower mentioned by M'Donald in his poem '*Alt an t-siucair,*' with the English of which I am not acquainted " (Armstrong).

It is *not* mentioned in the poem referred to, but in " *Oran an't Samhraidh* "—The Summer Song.

" S'cùraidh faileadh do mhuineil
A chrios-Chù-Chulainn nan càrn !
Na d' chruinn bhabaidean riabhach,
Lòineach, fhad luirgneach, sgiamhach.
Na d' thuim ghiobagach, dreach mhìn,
Bharr-bhùidhe, chasurlaich, àird ;
Timcheall thulmanan diamhair
Ma'm bi 'm biadh-ionain a fàs. "—M'DONALD.

Sweetly scented thy wreath,
Meadow-sweet of the cairns !
In round brindled clusters,
And softly fringed tresses,
Beautiful, tall, and graceful,
Creamy flowered, ringleted, high ;
Around sheltered hillocks
Where the wood-sorrel grows.

Welsh : *llysiu'r forwyn,* the maiden's flower.

S. filipendula—Dropwort. Gaelic and Irish : *grealan*—probably from *greadh,* to prepare food.

" A *gread* na cuilm. "—OSSIAN.
Preparing the feast.

Linnæus informs us that, " in a scarcity of corn the tubers have been eaten by men instead of food." Or from *greach,* a nut. Welsh : *crogedyf,*—*crogi,* to suspend. The tuberous roots are suspended on filaments ; hence the names *filipendula* and *dropwort.*

[1] Cù chullin's belt. Cùchullin was the most famous champion of the Ulster militia in the old Milesian times. He lived at the dawn of the Christian era. He was so called from *Cu,* a hound, and *Ullin,* the name of the province. Many stories are still extant regarding him.

Geum rivale—Water avens.[1] Gaelic: *machall uisge;* in Irish: *macha,* a head, and *all,* all—*i.e.,* allhead—the flower being large in proportion to the plant. *Uisge,* water. It grows in moist places only.

G. urbanum—Common avens. Gaelic: *machall coille,—coille,* wood, where it generally grows.

Dryas octopetala—White dryas. Gaelic: *machall monaidh,* the large-flowered mountain plant. (The name was given by an old man in Killin from a specimen from Ben Lawers in 1870.)

Potentilla anserina—Silverweed, white tansy. Gaelic: *brisgean* (written also *briosglan, brislean*), from *briosg* or *brisg,* brittle. *Brisgean milis,* sweet bread. "The *brisgean,* or wild skirret, is a succulent root not unfrequently used by the poorer people in some parts of the Highlands for bread" (Armstrong).

The skirret (see *Sium sisarum*) is not native. The plant here alluded to is *Potentilla anserina. Bar bhrisgean,* the flower. Welsh: *torllwydd,* from *tori,* to break.

P. reptans — Cinquefoil. Gaelic: *meangach,* branched or twigged, — *meang,* a branch; because of its runners, its long leaf, and flower-stalks. *Cuig bhileach,* five-leaved. Irish: *cuig mhear Mhuire,* Mary's five fingers. Welsh: *blysiu'r pump,* same meaning.

P. tormentilla — Common potentil, or tormentil. Gaelic: *leanartach* (from *leanar,* passive of verb *lean,* to follow). So common on the hills that it seems to follow one everywhere. *Bàrr braonan-nan-con,* the dogs' briar bud. *Braonan fraoch* (*fraoch,* heather). *Braonan,* the bud of a briar (Armstrong). *Braonan bachlag,* the earth-nut (*Bunium flexuosum*) (M'Donald), from *braon,* a drop.

> "Min-fheur chaorach is *bàrra-bhraonan.*"—M'INTYRE.
> Soft sheep grass and the flower of the tormentil.

Irish: *neamhnaid,* a pearl (in Gaelic: *neònaid*). Welsh: *tresgl y moch.*

Comarum palustre—Marsh cinquefoil. Gaelic: *cuig bhileach uisge,* the water five-leaved plant.

Fragaria vesca — Wood strawberry. Gaelic: *subh* (or *sùth*)

[1] *Avens,* a river, from the Celtic *an.* Welsh: *avon.* Gaelic: *abhainn.* Many river names in Europe and Asia are derived from this root — *e.g.,* Rhenus, the Rhine — *reidh-an,* the placid water. Garumnus, Garonne— *garbh-an,* the rough water. Marne—*marbh-an,* the dead water. Seine, a contraction of *seimh-an,* the smooth water, &c.

thalmhain, the earth's sap, the earth's delight (from *sùbh* or *sùgh*, sap, juice; also delight, pleasure, joy, mirth) ; *thalmhain*, of the earth.

> " Theirig *subh-thalmhain* nam bruach."—M'DONALD.
> The wild strawberries of the bank are done.

Subhan laire, the ground sap ; *tlachd subh*, pleasant fruit.

> " *Subhain laire* s'faile ghroiseidean."—M'INTYRE.
> Wild strawberries and the odour of gooseberries.

Suthag, a strawberry or raspberry.

> " Gur deirge n'ant *suthag* an ruthodh tha'd ghruidh."
> Thy cheeks are ruddier than the strawberry.

Irish : *catog*, the strawberry bush. *Cath*, seeds (the seedy fruit). Welsh : *mefussen.*

Rubus (from *rub*, red in Celtic), in reference to the colour of the fruit in some species.

Rubus chamæmorus—Cloudberry. Gaelic : *oireag*, variously written,—*oighreag, foighreag, feireag.* Irish : *eireag* (from *eireachd*, beauty).

> " Breac le *feireagan* is cruin dearg ceann."—M'INTYRE.
> Checkered with cloudberries with round red heads.

" The cloudberry is the most grateful fruit gathered by the Scotch Highlanders " (Neill).

The badge of Clan M'Farlane.

Cruban-na saona, " the dwarf mountain bramble." (O'Reilly, Armstrong, and others). Probably this is another name for the cloudberry, but its peculiar and untranslatable name furnishes no certain clue to what plant it was formerly applied.

R. saxatilis—Stone bramble. Gaelic : *caora bad miann*, the berry of the desirable cluster. *Ruiteaga*, redness, a slight tinge of red.

R. idæus—Raspberry. Gaelic : *preas sùbh chraobh* (*craobh*, a tree, a sprout, a bud), the bush with sappy sprouts.

> " Fàile nan *sùth-chraobh*
> A's nan ròsann."—M'INTYRE.
> The odour of rasps and roses.

Welsh : *mafon*,—*maf*, what is clustering. Gaelic : *preas shùidheag*, the sappy bush. *Sùghag*, the fruit (from *sùgh*, juice, sap).

R. fruticosus—Common bramble. Irish and Gaelic : *dreas*, plural *dris*. Welsh : *dyrys*,—the root *rys*, entangle, with prefix *dy*, force, irritation. In Gaelic and Welsh the words *dris* and *drysien* are applied to the bramble and briar indiscriminately.

> " An *dreas* a fàs gu h-urar."—OSSIAN.
> The bramble (or briar) freshly growing.
> " Am fear theid san *droighionn* domh
> Theid me san *dris* dà."—PROVERB.
> If one pass through thorns to me,
> I'll pass through brambles (or briars) to him.

Grian mhuine, the thorn (bush) that basks in the sun. *Dris muine,—muine*, a thorn, prickle, sting. *Smear phreas* (Irish: *smeur*), the bush that smears; *smearag*, that which smears (the fruit). Welsh: *miar*, the bramble. (*Miar* or *meur* in Gaelic means a finger.) *Smearachd*, fingering, greasing, smearing. (Compare Dutch, *smeeren;* German, *schmieren*, to smear or daub.) *Dris-smear*, another combination of the preceding names.

This plant is the badge of the Clan M'Lean.

R. cæsius—Blue bramble; dewberry bush. Gaelic: *preas-nan-gorm dhearc*, the blueberry bush.

> " Bar gach tolmain fo bhrat *gòrm dhearc*."—M'DONALD.
> Every knoll under a mantle of blueberries (dewberries).

The blue bramble is the badge of the Clan M'Nab.

Rosa canina—Dog-rose. Gaelic: *coin ròs*, dogs' rose (*coin*, gen. plural of *cù*, a dog). Greek: χυ-ων. Latin: *canis*. Sanscrit: *cunas*. Irish: *cù*. Welsh: *ki*.

Gaelic: *coin droighionn*, dogs' thorn. *Earrdhreas* or *fearra-dhris, earrad*, armour; suggested by its being armed with prickles.

> " Mar *mhucaig* na *fearra-dhris*."—M'ELLAR.
> Like hips on the briar.

Preas-nam-mucaig, the hip-bush — from *muc* (Welsh: *moch*), a pig, from the fancied resemblance of the seeds to pigs, being bristly. Irish: *sgeach mhadra*, the dogs' haw or bush. Welsh: *merddrain*. Gaelic: *ròs*, rose; cultivated rose, *ròs gharaidh*.

> " Bé sid an sealladh eibhinn !
> Do bhruachan glè-*dhearg ròs*."
> That was a joyful sight !
> Thy banks so rosy red.

R. rubiginosa—Sweet-briar (*briar*, Gaelic : a bodkin or pin). Gaelic: *dris chubhraidh*, the fragrant bramble. Irish: *sgeach-chumhra*, the fragrant haw or bush. *Cuirdris*, the twisting briar, —*cuir*, gen. sing. of *car*, to twist or wind.

Agrimonia eupatoria—Agrimony. Gaelic: *mur-dhraidhean*, —*mur*, sorrow, grief, affliction ; *draidhean*, another form of *dhroighionn* (see *Prunus spinosa*). *Draidh*, or *druidh*, also means a magician, which may refer to its supposed magical effects on troubles as well as diseases. A noted plant in olden times for the cure of various complaints. Irish : *marbh dhroighionn*,— *marbh-dhruidh*, a necromancer, or magician. *Geur bhileach*,— *geur*, sharp, sour, rigid ; *bhileach*, leaved ;—on account of its leaves being sharply serrated, or because of its bitter taste. *Mirean nam magh*, the merry one of the field. Welsh : *y dorllwyd*, the way to good luck.

Sanguisorba—Burnet. *A bhileach losgain*. The leaves good for burns and inflammations (*losgadh*, burning).

Alchemilla vulgaris—Common Lady's Mantle. Gaelic: *copan an druichd*, the dew-cup ; *falluing mhuire*, Mary's mantle. Irish: *dhearna mhuire*, Mary's palm. Gaelic: *crub leomhainn*, lion's paw ; *cota preasach nighean an righ*, the princesses' plaited garment. Irish : *leathach bhuidhe* (*leathach*, divided).

Alchemilla alpina—Alpine Lady's Mantle. Gaelic : *trusgan*, mantle. The satiny under-side of the leaves of this and the other species has given rise to the names *trusgan, falluing, cota*, and the English name, Lady's Mantle.

> " Tha *trusgan* faoilidh air cruit an aonich."—M'INTYRE.
> The mantle-grass on the ridge of the mountain.

The hills about Coire-cheathaich and Ben Doran (the district described by the poet) are covered with this beautiful plant. The word *trusgan*, mantle, may be used in this instance in its poetic sense.

Mespilus germanica—Medlar. Gaelic : *cran meidil* (M'Donald), said to be a corruption of Mespilus. Greek : μεσος, half, and πιλος, a bullet. The fruit resembles half a bullet.

Cratægus oxyacantha—Whitethorn, hawthorn. Gaelic : *sgith-each geal, drioghionn geal* (see *Prunus spinosa*), *geal*, white ; *preas nan sgeachag; sgeach*, a haw. Welsh : *draenen wen*, white thorn.

> " Mìòs bog nan ubhlan breac-mheallachd !
> Gu peurach plumbach *sgeachagach*,
> A' luisreadh sios le dearcagaibh,
> Cir, mhealach, beachach, groiseideach."—M'LACHUINN.
>
> Soft month of the spotted bossy apples !
> Producing pears, plums, and haws,
> Abounding in berries, wax,
> Honey, wasps, and gooseberries.

Uath or *huath* — the ancient Gaelic and Irish name — has several significations; but the root seems to be *hu* (Celtic), that which pervades. Welsh: *huad*, that which smells or has a scent (*huadgu*, a hound that scents). "The name hawthorn is supposed to be a corruption of the Dutch *hoeg*, a hedge-thorn. Although the fruit is generally called a haw, that name is derived from the tree which produces it, and does not, as is frequently supposed, take its name from the fruit it bears."—Jones. Hawthorn may only be a corruption of *huad-dracn*, scented thorns. The badge of the Clan Ogilvie.

Pyrus (from *peren*, Celtic for pear). Latin: *pyrum*. Armoric: *pèr*. Welsh: *peren*. French: *poire*.

Pyrus communis—Wild pear. Gaelic: *craobh pheurain fiadhain* (*peur*, the fruit), the wild pear-tree.

Pyrus malus—"*Mel* or *mal*, Celtic for the apple, which the Greeks have rendered μηλον, and the Latins *malus*."—Don. Welsh: *afal*. Anglo-Saxon: *æpl*. Norse: *apal*, apple. Gaelic: *ubhal; craobh ubhal fiadhain*, the wild apple-tree.

> " Do mheasan milis cubhraidh
> Nan *ubhlan* 's 'nam *peur*."—M'DONALD.
>
> Thy sweet and fragrant fruits,
> Apples and pears.

The old form of the word was *adhul* or *abhul*. The culture of apples must have been largely carried on in the Highlands in olden times, as appears from lines by Merlin, who flourished in A.D. 470, of which the following is a translation:—

"Sweet apple-tree loaded with the sweetest fruit, growing in the lonely wilds of the woods of Celyddon (Dunkeld), all seek thee for the sake of thy produce, but in vain; until Cadwaldr comes to the conference of the ford of Rheon, and Conan advances to oppose the Saxons in their career."

This poem is given under the name of *Afallanau*, or Orchard, by which Merlin perhaps means Athol—*i e., Abhal* or *Adhul*—which is believed by etymologists to acquire its name from its fruitfulness in apple-trees. *Goirteag* (from *goirt*, bitter), the sour or bitter one (the crab-apple). *Cuairtagan* (the fruit); *cuairt*, round, the roundies. Irish: *cueirt*.

> " 'San m'an Ruadh-aisrigh ah'fhas na *cuairtagan*."—M'INTYRE.
> It was near the red path where the crab-apples grew.

This plant is the badge of the Clan Lamont.

Pyrus aucuparia—Mountain-ash, rowan-tree. Old Irish and Gaelic: *luis*, drink (*luisreog*, a charm). The Highlanders formerly used to distil the fruit into a very good spirit. They also believed " that any part of this tree carried about with them would prove a sovereign charm against all the dire effects of enchantment or witchcraft."—Lightfoot (1772). *Fuinseag coille*, the wood enchantress, or the wood-ash (see *Circæa*); *craobh chaoran*, the berry-tree (*caor*, a berry). Irish: *pairtainn dearg*, the red crab.

> " Bu dh'eirge a ghruidh na *caoran*."—OSSIAN.
> His cheeks were ruddier than the rowan.

> " Sùil chorrach mar an dearcag,
> Fo rosg a dh-iathas dlù,
> Gruidhean mar na *caoran*
> Fo n' aodann tha leam cùin."—AN CAILIN DILEAS DONN.

> Thine eyes are like the blaeberry,
> Full and fresh upon the brae,
> Thy cheeks shall blush like the rowans
> On a mellow autumn day.
> (Translated by Professor J. S. Blackie.)

This plant is the badge of the Clan M'Lachlan.

Pyrus cydonia—Quince-tree. Gaelic: *craobh chuinnse*, corruption of quince, from French *coignassa*, pear-quince. Originally from Cydon in Candia.

AURANTIACEÆ.

Citrus aurantium—The orange. Gaelic: *òr ubhal*, golden apple; *òr mheas*, golden fruit; *òraisd*,[1] from Latin *aurum*. Irish: *or*. Welsh: *oyr*, gold.

> " 'S Phœbus dàth na'n tonn
> Air fiamh *òrensin*."—M'DONALD.
> And Phœbus colouring the waves
> With an orange tint.

Citrus medica—Citron. Gaelic: *craobh shitroin*.

Citrus limonum—Lemon. Gaelic: *crann limoin*. French: *limon*. Italian: *limone*.

---

[1] Spelt by M'Donald properly *orainis*. His spelling generally is far from correct, and the same word often spelt different ways. He is also much given to translating a name from the English.—Fergusson.

MYRTACEÆ.

Punica granatum—Pomegranate. Gaelic : *gràn ubhal (gràn,* Latin, *granum*), grain-apple.

"Tha do ghcuga mar lios *gràn ubhlan,* leis a'mheas a's taitniche."—SONG OF SOLOMON.

Thy plants are an orchard of pomegranates with pleasant fruits.

(Now generally written *pomgranat* in recent editions.)

Myrtus communis—Myrtle. Gaelic : *miortal.*

"An ait droighne fàsaidh an guithas, agus an ait drise fàsaidh am *miortal.*"
—ISAIAH lv. 13.

Instead of the thorn shall grow the fir, and instead of the briar, the myrtle.

ONAGRACEÆ.

Epilobium montanum—Mountain willow-herb. Gaelic : *an seileachan,* diminutive of *seileach* (Latin : *salix,* a willow), from the resemblance of its leaves to the willow. Welsh : *helyglys,* same meaning.

E. angustifolium — Rosebay. Gaelic : *seileachan frangach,* French willow. *Feamainn* (in Breadalbane), a common name for plants growing near water, especially if they have long stalks.

Circæa lutetiana and **alpina** — Enchantress's nightshade. Gaelic and Irish : *fuinnseach.* Not improbably from Irish *uinnseach,* playing the wanton—the reference being to the fruit, which lays hold of the clothes of passengers, from being covered with hooked prickles (as Circe is fabled to have done with her enchantments) ; or *fuinn,* a veil, a covering. The genus grows in shady places, where shrubs fit for incantations may be found. "*Fuinn* (a word of various significations), also means the earth ; and *seach,* dry — *i.e.,* the earth-dryer. *Fuinnseagal* (another Irish name), from *seagal* (Latin, *secale*), rye—*i.e.,* ground-rye" (Brockie). *Lus na h'oidhnan,* the maiden's or enchantress's weed.

LYTHRACEÆ.

Lythrum salicaria — Spiked lythrum, purple loosestrife. Gaelic : *lus an sith chainnt,* the peace-speaking plant.

"Chuir Dia oirnn *craobh sìth chainnt,*
Bha da'r dionadh gu leoir."—IAN LOM.

God put the peace-speaking plant over us,
Which sheltered us completely.

The name also applies to the common loosestrife, suggested probably by the Greek λυσις μαχη, of which the English name

D

"loosestrife" is a translation. Irish: *breallan leana*. *Breal*, a knob, a gland. It was employed as a remedy for glandular diseases, or from the appearance of the plant when in seed. *Breallan* means also a vessel. The capsule is enclosed in the tube of the calyx, as if it were in a vessel. *Lean*, a swamp. Generally growing in watery places.

HALORAGEÆ.

Myriophyllum spicatum and **alterniflorum.**—Water - milfoil. Gaelic and Irish: *snaithe bhathcadh* (from *snaith*, a thread, a filament; and *bàth*, drown), the drowned thread.

GROSSULARIACEÆ.

Ribes, said to be the name of an acid plant. (*Rhèum 'ribes*, mentioned by the Arabian physicians, a different plant). More probably from the Celtic *riob, rib*, or *reub*, to ensnare or entangle, to tear—many of the species being prickly. Latin: *ribes*. Gaelic: *spiontag*, currant, gooseberry. Irish: *spiontog, spin*. Latin: *spina*, a thorn; also *spion*, pull, pluck, tear away. Welsh: *yspinem*.

Ribes nigrum—Black currant. Gaelic: *raosar dubh*, the black currant. *Raosar* (Scotch, *rizzar*—from French, *raisin ;* Welsh, *rhyfion ;* Old English, *raisin tree*), for red currant.

R. rubrum—Red or white currants. Gaelic: *raosar dearg* or *geal*, red or white currants; *dearc frangach*, French berry.

R. grossularia — Gooseberry-bush. Gaelic: *preas ghrosaid* (written also *groseag, grosaid*), the gooseberry—from *grossulus*, diminutive of *grossus*, an unripe fig, — "so called because its berries resemble little half-ripe figs, *grossi*" (Loudon). French: *groseille*. Welsh: *grwysen*. Scotch: *grozet, grozel*.

"Suthan-lair's faile *ghroseidean*."—M'INTYRE.

Wild strawberry and the odour of gooseberries.

CRASSULACEÆ.

(From Latin, *crassus*, thick—in reference to the fleshy leaves and stem. Gaelic: *crasag*, corpulent.)

Sedum rhodiola—Rose-root. Gaelic and Irish: *lus nan laoch*, the heroes' plant; *laoch*, from the Irish, meaning a hero, a champion, a term of approbation for a young man.

The badge of the Clan Gunn.

S. acre—Stonecrop, wall-pepper. Gaelic and Irish: *grafan nan clach*, the stone's pickaxe. Welsh: *flyddarlys*, prick madam.

Also in Gaelic: *glas-lann* and *glas lean*, a green spot. Welsh: *manion y cerg*.

S. telephium—Orpine. Scotch: *orpie*. Gaelic: *orp* (from the French, *orpin*). *Lus nan laogh*, the calf or fawn's plant; *laogh*, a calf, a fawn, or young deer, a term of endearment for a young child. Irish: *laogh*. Welsh: *lho*. Manx: *leigh*. Armoric: *lue*. Welsh: *telefin* (from Latin, *telephium*).

Sempervirum tectorum — House-leek. Gaelic: *lus nan cluas*,[1] the ear-plant (the juice of the plant applied by itself, or mixed with cream, is used as a remedy for the ear-ache); *lus gharaidh*, the garden-wort; *oirp*, sometimes written *norp* (French, *orpin*); *tin gealach*, *tineas na gealaich*, lunacy—*tinn*, sick, and *gealach*, the moon (*geal*, white, from Greek, γαλα, milk);—it being employed as a remedy for various diseases, particularly those of women and children, and head complaints. Irish: *sinicin*, the little round hill; *tir-pin*, the ground-pine. Welsh: *llysie pen-ty*, house-top plant.

Cotyledon umbilicus—Navel-wort, wall-pennywort. Gaelic: *lamhag cat leacain*, the hill-cat's glove. Irish: *corn caisiol*, the wall drinking-horn (from *corn*, a cup, a convex surface; from its peltate round convex leaves). Latin: *cornu*, a horn. Welsh: *corn*. French: *corne*; and *caisiol*, a wall (or any stone building), where it frequently grows.

SAXIFRAGACEÆ.

Saxifraga — Saxifrage. Gaelic: *cloch-bhriseach* (Armstrong), stone-breaker—on account of its supposed medical virtue for that disease. Welsh: *cromil yr englyn*.

S. granulata—Meadow saxifrage. Gaelic and Irish: *moran*, which means many, a large number—probably referring to its many granular roots.

Chryosplenium oppositifolium — Golden saxifrage. Gaelic: *lus nan laogh* (the same for *Sedum telephium*). Irish: *clabrus*, from *clabar*, mud, growing in muddy places; *gloiris*, from *gloire*, glory, radiance,—another name given by the authorities for the "golden saxifrage;" but they probably mean *Saxifraga aizoides*, a more handsome plant, and extremely common beside the brooks and rivulets among the hills.

---

[1] This is what I always heard it called; but M'Donald gives *norn*, and in the Highland Society's Dictionary it is given *creamh-garaidh*, evidently a translation by the compilers, as they give the same name to the Leek.— FERGUSSON.

Parnassia palustris — Grass of Parnassus. Shaw gives the name *fionnsgoth* (*fionn*, white, pleasant, and *sgoth*, a flower), "a flower," but he does not specify which. *Finonan geal* has also been given as the name in certain districts, which seems to indicate that *fionnsgoth* is the true Celtic name.

ARALIACEÆ.

Hedera—"Has been derived from *hedra*, a cord, in Celtic" (Loudon).

Hedera helix—Ivy. Gaelic: *eidheann*, that which clothes or covers (from *eid*, to clothe, to cover); written also *eigheann* (*eige*, a web), *eidhne*, *eitheann*.

> " Spionn an *eitheann* o'craobh."—OLD POEM.
>
> Tear the ivy from the tree.
>
> " *Eitheann* nan crag."—OSSIAN.
>
> The rock-ivy.
>
> " Briseadh tro chreag nan eidheann dlu'
> Am fuaran ùr le torraman trom."—MIANN A BHARD AOSDA.
>
> Let the new-born gurgling fountain gush from the ivy-covered rock.

Faithleadgh, Irish: *faithlah*, that which takes hold or possession. Welsh: *iiddew* (from *eiddiaw*, to appropriate). Irish: *aighneann* (from *aighne*, affection), that which is symbolic of affection, from its clinging habit. *Gort*, sour, bitter—the berries being unpalatable to human beings, though eaten by birds. *Ialluin* (from *iall*, a thong, or that which surrounds); perhaps from the same root as *helix*. Greek: ἐιλέω (*eileo*, to encompass); also *iadh-shlat*, the twig that surrounds,—a name likewise given to the honeysuckle (*Lonicera periclymenum*), because it twines like the ivy—

> " Mar *iadh-shlat* ri stoc aosda."
>
> Like an ivy to an old trunk.

An gàth, a spear, a dart.
The badge of the Clan Gordon.

CORNACEÆ.

Cornus (from Latin: *cornu*, a horn). Gaelic: *corn*. French: *corne*. "The wood being thought to be hard and durable as horn."

Cornus sanguinea—Dogwood, cornel-tree. Gaelic: *coin-bhil*, dogwood; *conbhuiscne*, dog-tree (*baiscne*, Irish, a tree). Irish: *crann coirnel*, cornel-tree.

C. suecica—Dwarf cornel,—literally, Swedish cornel. Gaelic and Irish: *lus-a-chraois*, plant of gluttony (*craos*, a wide mouth; gluttony, appetite). " The berries have a sweet, waterish taste, and are supposed by the Highlanders to create a great appetite,—whence the Erse name of the plant" (Stuart of Killin).

UMBELLIFERÆ.

Hydrocotyle vulgaris—Marsh pennywort. Gaelic: *lus na peighinn*, the pennywort. Irish: *lus na pinghine* (O'Reilly), from the resemblance of its peltate leaf to a *peighinn*,—a Scotch penny, or the fourth part of a shilling sterling.

Eryngium maritimum—Sea-holly. Gaelic and Irish: *cuileann trágha*, sea-shore holly. (See *Ilex aquifolium*). Welsh: *y môr gelyn*, sea-holly (*celynen*, holly).

Sanicula europæa—Wood sanicle. Gaelic: *bodan coille*, wood-tail,—the little old man of the wood. Irish: *caogma*,—*caog*, to wink. *Buine*, an ulcer,—a noted herb, " to heal all green wounds speedily, or any ulcers. This is one of *Venus*, her herbs, to cure either wounds or what other mischief *Mars* inflicteth upon the body of man" (Culpepper). Welsh: *clust yr arth*, bear's-ear.

Conium maculatum—Hemlock. Gaelic: *minmhear* (Shaw), —smooth or small fingered, or branched, in reference to its foliage; *mongach mhear*, and *muinmhear*, — *mong* and *muing*, a mane, from its smooth, glossy, pinnatifid leaves. *Minbhar*, soft-topped or soft-foliaged. *Iteodha, iteotha,*—*ite*, feathers, plumage. The appearance of the foliage has evidently suggested these names, and not the qualities of the plant, although it is looked upon still with much antipathy.

> " Is coslach e measg chaich
> Ri *iteodha* an garadh."—M‘INTYRE.

Among other people he is like a hemlock in a garden.

" Mar so tha breitheanas a' fàs a nìos, mar an *iteotha* ann claisibh na machrach."— Hos. x. 4.

Thus judgment springeth up like a *hemlock* in the furrows of the field.

Welsh: *gwin dillad*, pain-killer. Manx: *aghue.*

> " Ta'n *aghue* veg shuyr da'n *aghue* vooar."—MANX PROVERB.

The little hemlock is sister to the big hemlock.

(A small sin is akin to the great one.)

Cicuta virosa — Water-hemlock. "The hemlock given to prisoners as poison" (Pliny); and that with which Socrates was poisoned. Gaelic and Irish: *fealla bog*, the soft deceiver; *feall*, treason, falsehood; and *fcallair* (*feall fhear*), a deceiver,—from the some root (Latin, *fallo*, to deceive). Welsh : *cegid*. Latin : *cicuta*.

Smyrnium olusatrum — Alexanders. Gaelic : *lus nan gràn dubh*, the plant with black seeds,—on account of its large black seeds. It was formerly eaten as a salad or pot-herb, whence, and from its blackness, the name *olusatrum* (Latin : *olus*, a vegetable, and *ater*, black). "'Alexanders,' because it was supposed to have been brought from Alexandria " (Ray).

Apium (from Celtic root, *abh*, a fluid or water, Latinised into *apium*).

Apium graveolens — Smallage, wild celery. Gaelic : *lus na smalaig*, a corruption of smallage. *Pearsal mhor*, the large parsley. Irish : *meirse*. Greek : μειρα, to divide ; or Anglo-Saxon : *merse*, a lake, sea. Latin : *mare*,—marshy ground being its habitat. Welsh : *persli frengig*, French parsley.

Petroselinum sativum—Parsley. Gaelic : *pearsal* (corruption from the Greek, πετρος, *petros*, a rock, and σελινον, *selinon*, parsley). *Muinean Mhuire*, Mary's sprouts. Welsh : *persli*.

Heliosciadium inundatum — Marshwort. Gaelic : *fualactar* (from *fual*, water). The plant grows in ditches, among water.

Carum carui—Caraway. Scotch : *carvie;* Gaelic : *carbhaidh* (a corruption from the generic name), from Caria, in Asia Minor, because it was originally found there ;—also written *carbhinn*.

"Cathair thalmhanta's *carbhinn* chroc cheannach."—M'INTYRE.
The yarrow and the horny-headed caraway.

Lus Mhic Chuimein, M'Cumin's wort. The name is derived from the Arabic *gamoùn*, the seeds of the plant *Cuminum cyminum* (*cumin*), which are used like those of caraway.

Bunium flexuosum—The earth-nut. Gaelic: *braonan bhuachail*, the shepherd's drop (or nut); *braonan bachlaig* (Shaw); *cno thalmhainn*,—*cno*, a nut, *thalmhainn*, earth,—ploughed land, ground. (Hebrew : דּלִים, *tilim*, ridges, heaps ; דלם, *talam*, break, as into ridges or furrows,—heap up. Latin : *tellus*. Arabic : *tēl*). Irish : *caor thalmhainn*, earth-berry ; *coircaran muic*, pig-berries, or pig-nuts. *Cutharlan*, a plant with a bulbous root.

Fœniculum vulgare—Fennel. Gaelic : *lus an t'saiodh*, the hayweed. *Fineal*, from Latin, *fœnum*, hay,—the smell of the plant resembling that of hay. Irish : *fineal chumhthra* (*cumhra*, sweet, fragrant). Welsh : *ffenigl*.

Ligusticum, from Liguria, where one species is common.

Ligusticum scoticum—Lovage. Gaelic: *siunas*, from *sion*, a blast, a storm,—growing in exposed situations. In the Western Isles, where it is frequent on the rocks at the sea-side, it is sometimes eaten raw as a salad, or boiled as greens. **Levisticum officinale** [1]—Common lovage. Gaelic: *luibh an liugair*, the cajoler's weed. It was supposed to soothe patients subject to hysterics and other complaints. Irish: *lus an liagaire*, the physician's plant, from which the Gaelic name is a corruption. Welsh: *dulys*, the dusky plant.

Meum athamanticum—Meu, spignel, baldmoney. Gaelic: *muilceann*. Scotch: *micken*,—*muilceann*,[2] possibly from *muil*, a scent; *muleideachd*, a bad smell (Shaw); *ceann*, a head or top. The whole plant is highly aromatic, with a hot flavour like lovage. Highlanders are very fond of chewing its roots.

Angelica—(So named from the supposed angelic virtues of some of the species).

A. sylvestris—Wood angelica. Gaelic: *lus nam buadha*, the plant having virtues or powers. *Cuinneog mhighe*, the whey bucket. *Galluran*, perhaps from *gall* (Greek: *gala*), milk, from its power of curdling milk; for this reason, hay containing it is considered unsuitable for cattle. Irish: *contran*. *Aingealag*: angelica.

Crithmum maritimum—Samphire. Gaelic: *saimbhir*, a corruption of the French name St Pierre (St Peter), from Greek, πέτρα, a rock or crag. (The samphire grows on cliffs on the shore). Gaelic: *lus nan cnàmh*, the digesting weed; *cnàmh* (from Greek: χναω; Welsh: *cnoi*; Irish: *cnaòi*), chew, digest. The herb makes a good salad, and is used medicinally. Irish: *grioloigin*,—*griol*, to slap, to strike.

Peucedanum ostruthium—Great masterwort. Gaelic: *mòr fhliodh* (Armstrong), the large excrescence, or the large chickweed.

P. officinale—Hog-fennel or sow-fennel. Gaelic: *fineal sraide* (Shaw),—*sraide*, a lane, a walk, a street. This plant is not found in Scotland, but was cultivated in olden times for the stimulating qualities attributed to the root.

---

[1] Levisticum, from Latin, *levo*, I assuage.

[2] In Invernesshire, *bricin* or *bricin dubh*, perhaps from *bri*, juice; or, as mentioned in Lightfoot, vol. i. p. 158, as Sibbald says it grows on the banks of the Breick Water in West Lothian, may not some native of the banks of the Breick have given it this local name in remembrance of seeing it growing on the banks of his native Breick?—FERGUSSON.

Anethum graveolens — Strong - scented or common dill. Gaelic and Irish : *dile* (M'Donald) (Latin : *diligo*), — *dile*, a word in Gaelic meaning love, affection, friendship. The whole plant is very aromatic, and is used for medicinal preparations.

Sium (from *siu*, "water in Celtic," Loudon), perhaps from *sjo* (Gothic), water, lake, sea.

S. sisarum—Skirrets. Gaelic : *crumagan* (Shaw), from *crom*, bent, crooked, from the form of its tubers. The tubers were boiled and served up with butter, and were declared by Worlridge, in 1682, to be "the sweetest, whitest, and most pleasant of roots;" formerly cultivated in Scotland under the name of "crummock," a corruption of the Gaelic name.

S. angustifolium—Water-parsnip. Gaelic : *folachdan* (Armstrong), from *folachd*, luxuriant vegetation ; *an*, water. Irish : *cosadh dubhadh*, the great water-parsnip (O'Reilly), (*cos*, a foot, stalk, shaft, and *dubh*, great, prodigious).

Pastinaca sativa — Parsnip. Gaelic : *meacan-an-righ*, the king's root, royal root. *Curran geal* (from *cur*, to sow, *geal*, white). Irish : *cuiridin ban*, the same meaning (*cuirim*, I plant or sow). Welsh : *moron gwynion*, field-carrot.

Ægopodium podagraria — Goat-, gout-, or bishop - weed. Gaelic : *lus an easbuig*,—*easbuig*, a bishop. A name also given to *Chrysanthemum leucanthemum*, but with a different signification.

Heracleum sphondylium — Cow-parsnip. Gaelic : *odharan*, from *odhar* (Greek : ωχρος; English : *ochre*), pale, dun, yellowish, in reference to the colour of the flower. *Meacan-a-chruidh*, the cow's plant. The plant is wholesome and nourishing for cattle. *Gunnachan sputachain*, squirt - guns. Children's name for the plant, because they make squirt-guns from its hollow stems.

Daucus carota—Carrot. Gaelic : *curran* (from *cur*, to sow), a root like that of the carrot. *Currait*, corruption from *carota*, which is said to be derived from the Celtic root *car*, red, from the colour of the root. *Muran*—(Welsh : *moron*), a plant with tapering roots. Irish : *curran bhuidhe*, the yellow root.

> "*Muran* brioghar 's an grunnasg lionmhar."—M'Intyre.
> The sappy carrot and the plentiful groundsel.

Irish : *mugoman*,—*mugan*, a mug, from the hollow bird's-nest-like flower.

Anthriscus { cerifolium, vulgaris, temulentum }—Chervil. Gaelic : *costag*, a common name for the chervils (from *cost*, an aromatic plant; Greek: κόστος, *kostos*, same meaning). *Costag a bhaile gheamhraidh* (*bhaile gheamhraidh*, cultivated ground). "*A. vulgaris* was formerly cultivated as a pot herb" (Dr Hooker).

Myrrhis (from Greek: μυρον, *myron*, perfume; Gaelic: *mirr*, —*tus agus mirr*, frankincense and myrrh).

M. odorata—Sweet cicely or great chervil. Gaelic: *cos uisge* (Shaw), the scented water-plant.[1] "Sweet chervil, gathered while young, and put among other herbs in a sallet, addeth a marvellous good relish to all the rest" (Parkinson).

Coriandrum (a name used by Pliny, derived from κορις, *coris*, a bug, from the fetid smell of the leaves).

C. sativum—Coriander. Gaelic: *coireiman,—lus a choire*, corruptions from the Greek. It is still used by druggists for various purposes, and by distillers for flavouring spirits.

LORANTHACEÆ.

Viscum album — Mistletoe. Gaelic and Irish: *uile-ice* (*uile*, Welsh: *hall* or *all*; Goth.: *alls*; German: *aller*; A. S.: *eal*; English: *all*; *ice*, Welsh: *iarc*, a cure or remedy), a nostrum, a panacea (M'Donald), all-heal. Armoric: *all-yiach*. Welsh: *oll-iach*. Irish: *uile iceach*. This is the ancient Druidical name for this plant. Pliny tells us, " The Druids (so they call their Magi) hold nothing in such sacred respect as the mistletoe, and the tree upon which it grows, provided it be an oak. ' Omnia sanantem appellantes suo vocabulo.' (They call it by a word signifying in their own language *All-heal*.) And having prepared sacrifices, and feast under the tree, they bring up two white bulls, whose horns are then first bound; the priest, in a white robe, ascends the tree, and cuts it off with a golden knife; it is received in a white sheet. Then, and not till then, they sacrifice the victims, praying that God would render His gift prosperous to those on whom He had bestowed it. When mistletoe is given as a potion, they are of opinion that it can remove animal barrenness, and that it is a remedy against all poisons." *Druidh-lus*, the Druid's weed. " The proper etymology is the ancient Celtic vocable *dru*, an oak, from which δρυς is taken " (Armstrong). *Sùgh dharaich*, the sap or substance of the oak, because it derives its substance from the oak, it being a parasite on that and other trees. (*Sùgh*, juice, sub-

---

[1] In Braemar it is commonly called *mirr.*—ED. 'Scottish Naturalist.'

E

stance, sap; Latin: *succus*). Irish: *guis*, viscous, sticky, on account of the sticky nature of the berries. French: *gui*.

CAPRIFOLIACEÆ.

Sambucus nigra — Common elder. Gaelic and Irish: *ruis*, meaning "wood." "The ancient name of the tree, which in the vulgar Irish is called *trom*" (O'Reilly); *druman* or *droman* (Sanscrit: *dru*, wood, tree; *drumas*, wood). Welsh: *ysgawen*, elder.

S. ebulus—Dwarf elder. Gaelic and Irish: *fliodh a bhalla*, the wall excrescence. *Mulart* "seems to be the same as the Welsh word *mwyllartaith* (*mwyll*, emollient, and *artaith*, torment")(Brockie). It was esteemed a powerful remedy for the innumerable ills that flesh is heir to. *Mulabhar* (*mul*, a multitude, and *bar*, top) may only be a corruption of *mulart*. The specific name is from ευβολη, *eubole*, an eruption. Welsh: *ysgawen Mair*, Mary's elder.

Viburnum opulus — Guelder-rose, Water-elder. Gaelic: *ćir-iocan*, heal-wax (Latin: *cera*; Greek: χηρυς; Welsh: *cwyr*, wax), the healing, wax like plant, from the waxy appearance of the flowers.

V. lantana—Wayfaring tree. Gaelic: *craobh fiadhain* (Armstrong), the wild or uncultivated tree.

Lonicera periclymenum—Woodbine, honeysuckle. Gaelic: *uillean* (elbows, arms, joints) elbow-like plant; *feith, feithlean.* Irish: *feathlog, fethlen*, from *feith*, a sinew, tendon, suggested by its twisting, sinewy stems. *Lus na meala*, the honey-plant, from *mil* (Greek: μελι; Latin: *mel*), honey. *Deolag*, or *deoghalag*, from *deothail*, to suck. Irish: *cas fa chrann*,[1] that which twists round the tree. *Baine gamhnach* (O'Reilly), the yearling's milk. A somewhat satirical name, implying that the sucking will produce scanty results. In Gaelic, *iadh shlat* is frequently applied both to this plant and to the ivy (see *Hedera helix*). Welsh: *gwyddfid*, tree-climber or hedge-climber.

RUBIACEÆ.

Rubia tinctorum—Madder. Gaelic: *madar* (Armstrong).
Galium aparine—Goose-grass; cleavers. Gaelic: *garbh lus*,

[1] In Strathardle and many other districts, *leum-a-chrann* (*leum*, jump, *crann*, a tree) alluding to its jumping or spreading from tree to tree. High. Soc. Dict. gives *duilliur-fèithlean*, probably from its darkening whatever grew under it.—FERGUSSON.

the rough weed. Irish: *airmeirg*, from *airm*, arms, weapons, from its stem being so profusely armed with retrograde prickles.

G. saxatile (Armstrong) — Heath bedstraw. *Madar fraoch*, heath madder. It grows abundantly among heather. O'Reilly gives this name also to *G. verum*.

G. verum—Yellow bedstraw. *Ruin*, *ruamh*, from *ruadh*, red. "The Highlanders use the roots to dye red colour. Their manner of doing so is this: The bark is stripped off the roots, in which bark the virtue principally lies. They then boil the roots thus stripped in water, to extract what little virtue remains in them; and after taking them out, they last of all put the bark into the liquor, and boil that and the yarn they intend to dye together, adding alum to fix the colour" (Lightfoot).

Lus an leasaich (in Glen Lyon) the rennet-weed. "The rennet is made as already mentioned, with the decoction of this herb. The Highlanders commonly added the leaves of the *Urtica dioica* or stinging-nettle, with a little salt" (Lightfoot). Irish: *baladh chnis* (O'Reilly), the scented form (*baladh*, odour, scent, *cneas*, form).

Asperula odorata—Woodruff. Gaelic: *lusa-caitheamh*.[1] Probably the Irish name *baladh chnis*, the scented form, is the woodruff, and not the lady's bedstraw; it is more appropriate to the former than to the latter.

VALERIANACEÆ.

Valeriana officinalis—Great wild valerian. Gaelic: *an tri-bhileach* (M'Kenzie); *lus na tri bhilean* (Armstrong), the three-leaved plant, from the pinnate leaves and an odd terminal one, forming three prominent leaflets. Irish: *lus na ttri ballan*, the plant with three teats (*ballan*, a teat); perhaps from its three prominent stamens (Brockie); *carthan curaigh* (*carthan*, useful, *curaigh*, a hero, a giant)—*i.e.*, the useful tall plant. Welsh: *y llysiewyn*, the beautiful plant; *y dri-aglog* (*dri*, three, *aglog*, burning; from its hot bitter taste).

V. dioica—Marsh or dwarf valerian. Irish: *carthan arraigh*, from *arrach*, dwarf; *caoirin leana*, that which gleams in the marsh (*caoir*, gleams, sparks, flames, flashes; *leana*, a swamp, a marsh). Although this plant is not recorded from Ireland, yet the names only occur in the Irish Gaelic.

---

[1] *Lusa-caitheamh*, the consumption herb, as it was much used for that disease.—FERGUSSON.

DIPSACEÆ.

Dipsacus sylvestris } Teasel,
      ,,   **fullonum** } Teasel, or fuller's teasel. Gaelic:
leadan,—liodan; liodan an fhucadair (*leadan* or *liodan*, a head
of hair, *fucadair*, a fuller of cloth); used for raising the nap
upon woollen cloth, by means of the hooked scales upon the
heads of the fuller's teasel. Irish : *taga*. Welsh : *llysie y cribef*,
carding plant, from *crib*, a comb, card.

Scabiosa succisa — Devil's bit scabious. Gaelic and Irish :
ura bhallach (*ur*, fresh, new ; *ballach*, from *ball*, a globular body,
from its globular-shaped flower-heads, or *ballach*, spotted. This
old Celtic word is found in many languages. Greek : βαλλω.
German : *ball.*) *Urach mhullaich*, bottle-topped (*urach*, a bottle,
from the form of the flower-head ; *mullach*, top). *Odharach
mhullaich*, a corruption of *urach*. (*Odhar* means dun or yellowish, but the flower is blue). *Greim an diabhail* (O'Reilly),
devil's bit, from its præmorse root, the roots appearing as if
bitten off. According to the old superstition, the devil, envy-
ing the benefits this plant might confer on mankind, bit away
a part of the root, hence the name. Welsh : *y glafrllys*, from
clafr, clawr, scab, mange, itch ; translation of *scabiosa*, from
scabies, the itch, which disorder it is said to cure.

Knautia arvensis—Corn-field knautia (so named in honour
of C. Knaut, a German botanist) or field scabious. Gaelic : *gille
guirmein*, the blue lad. Irish : *caba deasain*, the elegant cap ;
caba, a cap or hood ; and *deas*, neat, pretty, elegant. *Bodach
gorm*, the blue old man.

COMPOSITÆ.

Helminthia echioides — Ox tongue. Gaelic : *boglus* (Armstrong), a corruption from the Irish ; *bolglus*, ox-weed, from *bolg*,
a cow, an ox. A name also given to *Lycopsis arvensis*. *Bog
luibh*, same meaning.

Lactuca sativa—Lettuce. Gaelic and Irish : *liatus*, lettuce,
a corruption from *lactuca* (Latin : *lac*, milk), on account of the
milky sap which flows copiously when the plant is cut ; *luibh
inite*, the eatable plant. Irish : *billeog math*, the good leaf.
Welsh : *gwylath, gwyfluid, lath*, milk.

Sonchus oleraceus — Common sow - thistle, milk - thistle.
Gaelic and Irish : *bog ghioghan*, the soft thistle. Irish : *giogan*,
a thistle. *Baine muic*, sow's milk.

S. arvensis—Gaelic: *blioch fochain*, the corn milk-plant; *blioch*, milky; *fochan*, young corn. Welsh: *llaeth ysgallen*, milk-thistle (*ysgallen*, a thistle).

Hieracium pilosella—Mouse-ear hawkweed. Gaelic: *cluas luch*, mouse-ear; *cluas liath*, the grey ear.

H. murorum—Wall hawkweed. Irish: *srubhan na muc*, the pig's snout (*srubh*, a snout).

Taraxacum dens-leonis—Dandelion. Gaelic: *bearnan bride*.

> "Am *bearnan bride* s'a pheighinn rioghil."—M'INTYRE.
> The dandelion and the penny-royal.

Bearn, a notch, from its notched leaf; *bride*,[1] from *brigh*, sap, juice, with which the plant abounds; *bior nam bride* (*bior*, sharp, tooth-like); *fiacal leomhain*, lion's teeth. Welsh: *dant y llew*, the same meaning as dandelion (*dent de lion*) and *leontodon* (λεων, a lion; and οδους, a tooth), from the tooth like formation of the leaf. *Castearbhan nam muc* (Shaw)—The pig's sour-stemmed plant. Irish: *caisearbhan, cais-t'searbhain, castearbhan* (*cais*, a word of many significations, but here from *cas*, a foot; *caiseag*, the stem of a plant; *searbh*, bitter, sour).

Cichorium intybus—Succory or Chicory. Gaelic: *lus an t-suicair*, a corruption from *cichorium*, which was so named from the Egyptian word *chikouryeh*. Pliny remarks that the Egyptians made their chicory of much consequence, as it or a similar plant constituted half the food of the common people. It is also called in Gaelic *castearbhan*, the sour-stemmed plant.

C. endiva—Endive. Gaelic: *enach gharaidh* (*enach*, corruption of *endiva*, "from the Arabic name *hendibeh*" (Du Théis), *garadh*, a garden). Welsh: *ysgali y meirch*, horse-thistle.

Lapsana communis—Nipple-wort. Gaelic: *duilleag mhaith*, the good leaf; *duilleag mhìn*, the smooth leaf. Irish: *duilleog bhrighid*, the efficacious leaf, or perhaps St Bridget's leaf, the saint who, according to Celtic superstition, had the power of revealing to girls their future husbands. French: *herbe aux mamelles*, having been formerly applied to the breasts of women to allay irritation caused by nursing. *Duilleog bhraghad*, or *braighe*, the breast-leaf.

---

[1] "Most certainly *bride* comes from its being in flower plentifully on *latha fheill-brìde*."—FERGUSSON.
Bride is also a corruption of *Bbrighit*, St Bridget. *Latha Fheill-Brìghde*, Candlemas, St Bridget's Day.

"Tha do phòg mar ùbhlan garaidh,
'S tha do *bhraighe* mar an neoinean."—M'INTYRE, *Oran Gaoil.*
Thy kiss is like the apples of the garden,
And thy bosom like the daisy.

" If it was used by the French for rubbing the breasts, nothing seems more likely than that it would be also so used by the Celts of Ireland and Scotland, which would at once give it the name of *dulleog braghad*" (Fergusson).

Arctium—Celtic : *art*, a bear. Greek : αρκτος, from the rough bristly hair of the fruit.

A. lappa—Burdock. Gaelic and Irish : *suirichean suirich*, the foolish wooer (*suiriche*, a fool ; *suirich*, a lover or wooer); *seircean suirich*, affectionate wooer (*seirc*, affection). *Mac-an-dogha*,[1] the mischievous plant (*mac-an* for *meacan*, a plant) ; *doghadh*, mischievous (Shaw). *Meacan-tobhach-dubh*, the plant that seizes (*tobhach*, wrestling, seizing, inducing ; *dubh*, black, or large). *Leadan liosda* (*leadan*, a head of hair ; *liosda*, stiff). Irish : *copag tuaithil*, the ungainly docken ; *ceosan*, the bur, or fruit.

"Mar *cheosan* air sgiathan fhirein."—OSSIAN.
Like bur clinging to the eagle's wing.

Welsh : *cynghau*, closely packed. *Cribe y bleidd*, wolf's comb. *Caca muci*, puck's dung. *Lappa*, from Celtic, *llap* (Loudon). Gaelic (for hand) *làmh.* Welsh : *llamh.*

Carduus heterophyllus—Melancholy thistle. Gaelic : *cluas an fheidh*, the deer's ear.

C. palustris—Marsh-thistle. Gaelic : *cluaran leana* (*cluaran*, a thistle ; *lean*, a swamp) ;

"Lubadh *cluaran* mu Lora nan sion."—OSSIAN.
Bending the thistle round Lora of the storms.

Cluaran, a general name for all the thistles. Welsh : *ys gallen.*

C. lanceolatus—Spear-thistle. Gaelic : *an cluaran deilgneach*, the prickly thistle (*deilgne*, prickle-thorn).

C. arvensis—Corn-thistle. Gaelic : *aigheannach*, the valiant one (from *aighe*, stout, valiant).

C. marianus—Mary's thistle. Gaelic : *fothannan beannuichte.* Irish : *fothannan beanduighte* (Latin : *benedictus*), the blessed thistle (so called from the superstition that its leaves are stained with the Virgin Mary's milk) ; *fothannan, foghnan, fonndan*, a thistle. Danish : *fön*, thistle-down.

---

[1] *Dogha* also means burnt or singed. It was formerly burned to procure from its ashes a white alkaline salt, as good as the best potash.

"Leannaibh am *foghnan*."—OSSIAN.
Pursue the thistle-down.
" 'Feadh nan raointean lom ud,
Far nach cinn na *foth'nain*."
Among these bare hillsides,
Where the thistles will not grow.

M'Donald has another name, *cluaran òir*, the gold thistle.

" Gaoir bheachainn bhùi 's ruadha
Ri deoghladh *chluaran òir*."
The buzzing of yellow and red wasps
Sucking the golden thistle.

It is uncertain to which thistle, if any, the reference is made, unless it be to *Carlina vulgaris*, the carline thistle. *Cluaran*, occasionally means a daisy, *Chrysanthemum segetum*, one of its names being *liathan*.

" *Liath chluaran* nam magh."—OSSIAN.
The hoary thistle (or daisy) of the field.

Here the reference is evidently to the corn-marigold; in all probability M'Donald refers to the same flower, and not to any thistle (see *Chrysanthemum segetum*).

The thistle, the badge of the Clan Stewart.

Cynara scolymus—Artichoke. Gaelic: *farusgag*, from *farusg*, the inner rind, the part used being the lower part of the receptacle of the flower, freed from the bristles and seed-down, and the lower part of the leaves of the involucre. *Bliosan*, not unlikely to be a contraction from *bli-liosan*,—*bli* (*bligh*), milk (with its florets milk was formerly coagulated); and *lios*, a garden. These names apply also to *Helianthus tuberosus*, Jerusalem artichoke, especially to the tubers; and *plur na greine*, to the flower, from the popular error that the flower turns with the sun.

Centaurea nigra—Knapweed. Gaelic: *cnapan dubh*, the black knob (from *cnap*, a knob; Welsh, Armoric, and Irish: *cnap*; Saxon: *cnæp*; Danish *cnap*.) *Mullach dubh*, the black top. Irish: *niansgoth*, the daughter's flower (*nian*, a daughter; *sgoth*, a flower).

C. cyanus—Blue-bottle. Gaelic: *gorman*, the blue one. In some places, *gille-guirmean*, the blue lad. *Curachd chubhaig*, the cuckoo's cap or hood. Irish: *curac na cuig*, the same meaning. Welsh: *penlas wen*, blue headed beauty.

Artemisia vulgaris—Mugwort. Gaelic: *liath lus*, the grey weed. *Mòr manta* (Shaw), the large demure-looking plant (*mòr*, large; *manta*, demure, bashful). *Mughard*, Mugwort (*mugan*, in Irish, a *mug*, or *mugart*, a hog). Irish: *bofulan ban*, or *buafannan bàn*, the white toad, or serpent (*buaf*, a toad; *buafa*, a serpent; Latin: *bufo*, a toad); *buafannan liath*, the grey toad or serpent. Welsh: *llwydlys*, grey weed.

A absinthium—Common wormwood. Gaelic: *buramaide*. Irish: *borramotor*, also *burbun* (*burrais*, a worm or caterpillar; *maide*, wood)—*i. e.*, wormwood. *Searbh luibh*, bitter plant.

> "Chuir e air mhisg mə le *searbh-luibhean*."—STUART.
> He hath made me drunk with wormwood.

> "Mar a *bhurmaid*."
> Like the wormwood.

It was formerly used instead of hops to increase the intoxicating quality of malt liquor. *Roide*, gall, bitterness. *Graban* (from Gothic, *grub*, dig).[1] Welsh: *bermod chwerwlys*, bitter weed.

A. abrotanum — Southernwood. Gaelic: *meath chaltuinn*. (*Meath*, Latin *mitis*, faint, weary, effeminate. Its strong smell is said to prevent faintness and weariness. *Caltuinn*, from *càl*, Latin: *cald*; Italian: *cala*; French: *cale*, a bay, sea-shore, a harbour.) It grows in similar situations to *A. maritima*. Irish: *surabhan*, *suramont*, and Welsh: *siwdrmwt*. The sour one (*sùr*, sour), and "southernwood," also from the same root. Welsh: *llysier cyrff*, ale-wort (*cyrff*, Latin, *cervisia*, ale), it being frequently used instead of hops to give a bitter taste to malt liquors.

Gnaphalium dioicum, G. sylvaticum — Cudweed. Gaelic: *cat luibh*, the cat's weed. *Gnàbh*, or *cnàmh lus*, the weed that wastes slowly (from γναφάλιον), a word with which Dioscorides describes a plant with white soft leaves, which served the purpose of cotton. This well describes these plants. They have all beautifully soft woolly leaves; and, on account of the permanence of the form and colour of their dry flowers, are called "Everlasting."

Filago germanica—Common cotton rose. Gaelic and Irish: *liath lus roid*, the gall (or wormwood) grey weed.

---

[1] The occasional occurrence of Gothic roots in plants' names in the Western Highlands and Isles, is accounted for by the conquest of these parts by the Norwegians in the ninth century, and the fact of their rule existing there for at least two centuries under the sway of the Norwegian kings of Man and the Isles.

Petasites vulgaris—Butter-bur, pestilence-wort. Gaelic and Irish: *gallan mòr*, the big branch, possibly referring to its large leaf. Greek: γαλανος, mast. Danish: *galan*, a stripling. *Pobal*, more correctly *pubal.* Welsh: *pabel*, a tent, a covering.

> "Shidhich iad am *pubuill.*"—OSSIAN.
> They pitched their tents.

The Greek name, πετασος, a broad covering, in allusion to its large leaves, which are larger than that of any other British plant, and form an excellent shelter for small animals.

Tussilago farfara—Colt's-foot. Gaelic: *cluas liath*, grey ear; *gorm liath*, greyish green; *duilliur spuing*, the tinder-leaf.

> "Cho tioram ri *spuing.*"
> As dry as tinder.

The leaf, dipped in saltpetre and then dried, made excellent tinder or touchwood. Gaelic and Irish: *fathan* or *athan*, meaning fire. It was used for lighting fire. The leaves were smoked before the introduction of tobacco, and still form the principal ingredient in the British herb tobacco. *Gallan-greannchair* (*gallan*, see "Petasites;" *greann*, hair standing on end, a beard), probably referring to its pappus. Irish: *cassachdaighe* (O'Reilly), a remedy for a cough (*casachd*, a cough; *aighe* or *ice*, a remedy). "The leaves smoked, or a syrup or decoction of them and the flowers, stand recommended in coughs and other disorders of the breast and lungs" (Lightfoot). Welsh: *carn y ebol* (*carn*, hoof, and *ebol*, foal or colt), colt's-foot.

Senecio vulgaris—Groundsel. Gaelic: *am bualan*, from *bual*, a remedy. *Lus Phàra liath*,[1] grey Peter's-weed, a name suggested by its aged appearance, even in the spring-time. Latin: *senecio*. Welsh: *ben-felan*, sly woman. *Sail bhuinn* (*sail*, a heel; *buinn*, an ulcer). "The Highlanders use it externally in cataplasms as a cooler, and to bring on suppurations" (Lightfoot). *Grunnasg* (from *grunnd*, ground; German: *grund*). Welsh: *grunsel*.

> "Muran brioghar s'an *grunnasg* lionmhor."—M'INTYRE.
> The sappy carrot and the plentiful groundsel.

Irish: *crann lus*, the plough-weed. *Buafanan na h' easgaran*

---

[1] In Breadalbane, Glenlyon, and other places, the plant is called *Lus Phàra liath*—

> "*Lus Phàra liath* cuiridh e ghoimh as a chraimh."
> The groundsel will extinguish acute pain in the bone—

it being frequently applied as a cure for rheumatic pains.

(*buaf*, a toad, a serpent, but in this name evidently a corruption from *bualan*, a remedy, or *buadh*, to overcome ; *easgaran*, the plague), a remedy for the plague. A name given also to the ragwort.

S. Jacobæa—Ragwort. Gaelic and Irish : *buadhlan buidhe* (from *buadh*, to overcome ; *buidhe*, yellow) ; *buadhghallan*, the stripling or branch that overcomes ; *guiseag bhuidhe*, or *cuiseag*, the yellow - stalked plant ; *cuiseag*, a stalk. Welsh : *llysiu'r ysgyfarnog*, the hare's plant ; *llysiu'r nedir*, the serpent's weed —agreeing with one of its Irish names, *buafanan,—buaf*, a serpent or toad.

Inula Helenium — Elecampane, said to be from the officinal name, *inula campana*, but probably a corruption of Helénula, Little Helen (Jones). Greek : ἐλενος, the elecampane. Gaelic : *àillean*, from *àille*, beautiful, handsome. Irish : *Ellea* (Gaelic, *Eilidh*), Helen. The famous Helen of Troy, who is said to have availed herself of the cosmetic properties of the plant. *Creamh*, sometimes, but more generally applied to *Allium ursinum* (which see).

Bellis perennis—Daisy. Gaelic and Irish : *neòinan*, or *nòinean*, the noon-flower (from *noìn*, noon ; Welsh : *nawn* ; Latin : *nona*, the ninth hour, from *novem*, ninth. The ninth hour, or three in the afternoon, was the noon of the ancients).

> " 'San *nèoinean* beag's mo lamh air cluin."—MIAN A BHARD AOSDA.
> And the little daisy surrounding my hillock.

Buidheag (in Perthshire), the little yellow one.

> " Geibh sinn a *bhuidheag* san lòn."—OLD SONG.
> We shall find the daisy in the meadow.

Gugan (Armstrong), a daisy, a bud, a flower.

Chrysanthemum segetum — Corn - marigold. Gaelic : *bile buidhe*, the yellow blossom. *Bileach choigreach*, the stranger or foreigner. *Liathan*, Irish, *lia*, the hoary grey one (from Greek λεῖος ; Welsh : *llwyd*), on account of the light-grey appearance of the plant, expressed botanically by the term glaucous. *An dithean òir*, the golden flower, or chrysanthemum (χρυσος, gold ; ανθος, a flower).

> " Mar mhìn-chloch nan òr *dhithean* beag."
> Like the tender breast of the little marigold.

> " Do *dhithean* lurach luaineach
> Mar thuarneagan de'n 'òr."—M'DONALD.
> Thy lovely marigolds like waving cups of gold.

Dìthean is frequently used in a general sense for " flower," also for " darnel."

" Tir nan *dìthean* miadar daite. "
Land of flowers, meadow dyed.
" *Dìthein* nan gleann. "
The flowers of the valley.

Welsh : *gold mair,* marigold. Irish : *buafanan buidhe,* the yellow toad.

C. leucanthemum—Ox-eye. Gaelic : *an neònan mòr,* the big daisy. *Am brèinean-brothach* [1] (*brèine,* stench ; *brothach,* scabby). *Easbuigban,* from Irish *easbudh,* silly, idle (*easbudh brothach,* the King's-evil). This plant was esteemed an excellent remedy for that complaint. Irish : *easbuig speain* (*Speain* or *Easbain,* Spain).

Anthemis nobilis—Common chamomile. *Camomhil,* from the Greek χαμαι μηλος, which Pliny informs us was applied to the plant on account of its smelling like apples. (Spanish : *mancinilla,* a little apple.) *Lus-nan-cam-bhil* (M'Kenzie), the plant with drooping flowers. The plant is well distinguished by its flowers, which droop, or are *bent down,* before expansion ; but though the name is thus applicable, it is only a corruption from the Greek.

" Bi'dh mionntain, *camomhil* s'sobhraichean
Geur bhileach, lonach, luasganach."—M'INTYRE.
There will be mints, chamomile, and primroses,
Sharp-leaved, prattling, restless.

Luibh-leighis, the healing plant. This plant is held in considerable repute, both in the popular and scientific Materia Medica.

A. pyrethrum—Pellitory of Spain. Gaelic : *lus na Spàine,* the Spanish weed.

A. arvensis—Field chamomile. Irish : *coman mionla* (*coman,* a common ; *mionla,* fine-foliaged. Gaelic : *mìn lach*).

Matricaria inodora—Scentless May-weed. Gaelic : *buidheag an arbhair,* the corn daisy. *Camomhil feadhain,* wild chamomile. Welsh : *llygad yr ych,* ox-eye.

Tanacetum vulgare — Tansy. Gaelic : *lus na Fraing,* the French weed. (French, *tanaisie.*) Irish : *tamhsae,* corruptions from *Athanasia.* (Greek : a, privative, and θαvατος, death, *i.e.,* a plant which does not perish—a name far from applicable to this species).

Eupatorium cannabinum — Hemp agrimony. Gaelic and

---

[1] *Breinean-brothach* was probably also applied to *A. cotula,* for which there is no Gaelic name recorded.

Irish : *cnaib uisge* or *caineab uisge*, water-hemp (from Greek κανναβις ; Latin, *cannabis*, hemp ; the root *can*, white).

Bidens cernua — Bur marigold. Irish : *sceachog Mhuire*, Mary's haw.

Achillea ptarmica—Sneezewort. Gaelic : *cruaidh lus*, hard weed. (Latin : *crudus*, hard, inflexible). *Meacan ragaim*, the stiff plant. *Roibhe*, moppy.

A. millefolium—Yarrow. Gaelic : *lus chosgadh na fola*, the plant that stops bleeding. *Lus na fola*, the blood-weed. *Earr thalmhainn*, that which clothes the earth (*earr*, clothe, array). *Athair thalmhainn*, the ground father. *Cathair thalmhainn*, the ground seat or chair. Probably alterations of *earr* (for *thalmhainn* see *Bunium flexuosum*).

" *Cathair thalmhainn's* carbhin chroc-cheannach."—M'INTYRE.

The yarrow and the horny-headed caraway.

Solidago virgaurea—Golden rod. Gaelic : *fuinseag coille* ? A name given by Shaw to the herb called " *Virgo pastoris.*" Also one of the names of the mountain-ash (*Pyrus aucuparia*, which see).

Jasione montana—Sheep - bit. Gaelic : *dubhan nan caora* (O'Reilly). *Dubhan*, a kidney ; *caora*, sheep.

CAMPANULACEÆ.

Campanula—Gaelic : *barr-cluigeannach*, bell-flowered.

" Barr-cluigeannach-sinnteach gorm-bhileach."

Bell-flowered extended, blue-petalled.

C. rotundifolia—Round-leaved bell-flower. Gaelic : *bròg na cubhaig*, the cuckoo's shoe. *Am pluran cluigeannach*, the bell-like flower. Welsh : *bysedd ellyllon*, imp's fingers. Scotch : witch's thimbles.

Lobelia dortmanna—Water-lobelia. *Plùr an lochain*, the lake-flower.

ERICACEÆ.

Erica tetralix—Cross-leaved heath. Gaelic : *fraoch frangach*, French heath. *Fraoch an ruinnse*, rinsing heath ; a bunch of its stems tied together makes an excellent scouring-brush, the other kinds being too coarse. (*Fraoch*, anciently *fraech*.) Welsh : *grûg*. Greek : ἐρείκω, *ereiko*, to break, from the supposed quality of some of the species in breaking the stone (medicinally). The primary meaning seems to be to burst, to break, and appears to be cognate with the Latin, *fractum*. *Fraoch* also means wrath, fury, hunger. "*Laoch bu gharg fraoch*" (Ull.), a hero of the fiercest wrath. "*Fraoch!*" fury, the war-cry of the M'Donalds.

E. vagans—Cornish heath. Celtic: *gooneleg* (Dr Hooker), the bee's resort.

E. cinerea—Smooth-leaved heath. Gaelic: *fraoch bhadain*, the tufted heath.

> "Barr an *fhraoch bhadanaich*."—OLD SONG.
>
> The top of the tufted heath.

> "Gur badanach caoineil mileanta
> Cruinn mopach, mìn cruth, mongoineach
> Fraoch groganach, dù dhonn grìs dearg."—M'INTYRE.

Literally—

> That heath so tufty, mellow, sweet-lipped,
> Round, moppy, delicate, ruddy,
> Stumpy, brown, and purple.

Fraoch an dearrasain, the heath that makes a rustling or buzzing sound.

The badge of Clan M'Donald.

Calluna vulgaris—Ling heather. Gaelic: *fraoch*. Heath or heather is still applied to many important domestic purposes, thatching houses, &c., and "the hardy Highlanders frequently make their beds with it—the roots down and the tops upwards—and formerly tanned leather, dyed yarn, and even made a kind of ale from its tender tops." *Langa* (M'Kenzie), ling.

Arbutus uva-ursi—Red bearberry. Gaelic: *grainnseag*, small, grain-like. It has small red *berries*, which are a favourite food for moorfowl. *Braoileag nan con*, the dogs' berry.

A. alpina—The black bearberry. Gaelic: *grainnseag dhubh*, the black grain-like berry.

A. unedo—Strawberry-tree. Irish: *caithne* (O'Donovan). *Caithim*, I eat or consume.

Vaccinium myrtillus—Whortleberry. Gaelic: *lus nan dearc*, the berry plant (*dearc*,[1] a berry). *Gearr-dhearc*, sour berry. *Fraochan*, that which grows among the heather. The berries are used medicinally by the Highlanders, and made into tarts and jellies, which last is mixed with whisky to give it a relish for strangers. *Dearcan-fithich*, the raven's berries.

V. vitis-idæa—Cowberry; red whortleberry; cranberry.

---

[1] Originally from *dearc*, the eye; Sansk., *darç*, to see. The dark fruit resembling the pupil of the eye—hence the frequent comparisons of the eye (*sùil*) to this fruit (*dearcag*) in Gaelic poetry.

Gaelic : *lus nam broighleag.* Irish : *braighleog* (from *braigh*, top, summit, a mountain), the mountain-plant ; ordinary signification, a berry. *Bó-dhearc,* cowberry. ("*Bó,* a cow, from which the Greeks derived βοος, an ox"—Armstrong.) Latin : *vacca* and *vaccinium.*

> " Do leacan chaoimhneil gu *dearcach brioghleagach.*"
> Thy gentle slopes abounding with whortleberries and cowberries.

Badge of Clan M'Leod.

V. oxycoccos—Cranberry. Gaelic and Irish : *muileag,* a word meaning a little frog ; the frogberry. It flourishes best in boggy situations. *Fraochag,* because it grows among the heather. *Monog,* bog or peat berry. *Mionag,* the small berry.

V. uliginosum—The bogberry. Gaelic : *dearc roide,* the gall or bitter berry. The fruit abounds with an acid juice ; when the ripe fruit is eaten, it occasions headache and giddiness.

ILEACEÆ.

Ilex aquifolium—Holly. Gaelic and Irish : *cuileann.* Welsh : *celyn.* A.-S. : *holegn.* (*C* in Gaelic corresponds with *h* in the Germanic languages.) *Cùl,* guard, defence ; *cùil,* that which prohibits. Compare also *cuilg,* gen. of *colg,* a prickle, or any sharp-pointed thing. The lower leaves of this tree are very prickly, and thus guard against cattle eating the young shoots. Welsh : *celyn,* tree, shelterer or protector ; *cel,* conceal, shelter, cover.

> " Ma theid thu rùisgte troimh thom droighinn
> 'S coiseachd cas-lom air *preas cuileann*
> Cadal gun lein' air an eanntaig,
> 'S racadal itheadh gunn draing ort," &c.—BLAR SHUNADAIL.

> If you go naked through a thorn thicket,
> And walk barefooted on the *holly,*
> Sleep without a shirt on the nettle,
> And eat horse-radish without a grin, &c.

OLEACEÆ.

Olea europæa—European olive. Gaelic and Irish : *crann oladh* or *ola* (Greek : ελαια, a word, according to Du Théis, derived from the Celtic ; Welsh : *oleu*), the oil-tree.

> " Sgaoilidh e gheugan, agus bithidh a mhaise mar an *crann-oladh.*"
> " He will spread his branches, and his beauty shall be as the *olive tree.*"—
> HOSEA xiv. 6.

Syringa vulgaris—Lilac-tree. Gaelic : *craobh liath ghorm,* the lilac-tree.

Ligustrum vulgare—Privet. Gaelic : *ras chrann sìr uine,* the evergreen shrubbery-tree. *Priobaid* (M·Donald). (Latin : *privatus*; Irish : *priobhaid*, secrecy, privacy). Its chief use is to form hedges that are required for shelter, ornament, and privacy.

Fraxinus excelsior—Ash. Gael and Irish : *craobh uinnseann*. Irish : *uinscann, uimhseann*, altered into *fuinse, fuinsean, fuinseog*.

> " Gabhaidh an t' *uinnseann* as an àllt
> 'S a chàlltuinn as a phreas."—PROVERB.
>
> The *ash* will kindle out of the burn,
> And the hazel out of the bush.

Welsh : *onen*, corresponding to another Irish name, *nion*. Gaelic : *nuin*, and also *oinsean*. The names refer principally to the wood, and the primary idea seems to be lasting, long-continuing, *on* (in Welsh), that which is in continuity. *Uimh*, number; *seann*, ancient, old; *ùine*, time, season. *Nuin*, also the letter N. Heb., *nun*. *Fuinnseann* (see *Circœa*), though from the same root, may have been suggested by its frequent use in the charms and enchantments so common in olden times, especially against the bites of serpents, and the influence of the "Old Serpent." Pennant, in 1772, mentions : " In many parts of the Highlands, at the birth of a child, the nurse puts the end of a green stick of ash into the fire, and while it is burning, receives into a spoon the sap or juice which oozes out at the other end, and administers this to the new-born babe." Serpents were supposed to have a special horror of its leaves.

> "Theid an nathair troimh an teine dhearg
> Mu'n teid i troimh dhuilleach an ùinnsinn."

The serpent will go through fire, rather than through the leaves of the ash.

The same superstition was equally common in other countries, and the name "ash," which is said to be from the Celtic word *œsc*, a pike, is more likely to be from the word *asc*, a snake, an adder.[1] German : *die esche*.

The badge of Clan Menzies.

GENTIANACEÆ.

Gentiana campestris—Field gentian. Gaelic: *lus a chrùbain*, the crouching plant, or the plant good for the disease called *crùbain*, "which attacks cows, and is supposed to be produced

---

[1] In Scandinavian mythology the first man was called *Ask*, and the first woman *Ambla*—ash and elm. The court of the gods is represented in the Edda as held under an ash—*Yggdrasil*. Connected with these circumstances probably arose the superstitions.—CHAMBERS'S ENCYCLOPÆDIA.

by hard grass, scanty pasture, or other causes. The cows become lean and weak, with their hind-legs contracted towards the fore-feet, as if pulled by a rope" (Armstrong). This plant, in common with others of this genus, acts as an excellent tonic ; its qualities were well known in olden times. Welsh : *crwynllys*, bent-weed ; *cryn*, bend, curve. Gaelic : *creamh*, is given also as a name for gentian.

> " 'N creamh na charaichean,
> Am bac nan staidhrachean."—M'INTYRE.

Which Dr Armstrong translates, "gentian in beds or plots." The name *creamh* also applies to the leek. *Creamh*, hart's-tongue fern, garlic, and elecampane. Welsh : *craf*, garlic.

Erythræa, from ερυθρος, *erythros*, red flowers.

E. centaurium—Centaury ; red gentian. Irish : *ceadharlach* (O'Reilly), the *centaur*. It is said that with this plant Chiron cured the wound caused by the arrows of Hercules in the Centaur's foot. Gaelic, according to Armstrong : *ceud bhileach*, meaning hundred-leaved, a corruption of the Irish name (*Ceud*, Irish : *ceadh ;* Latin : *centum*, a hundred),—the origin of the name being probably misunderstood.

E. littoralis—Dwarf-tufted century. Gaelic and Irish : *dreim-ire muire*, the sea-side scrambler. *Dreim*, climb, clamber, scramble : *muire ;* Latin : *mare ;* German : *meer*, the sea.

Chlora perfoliata—Yellow-wort. Gaelic and Irish : *dreimire buidhe*, the yellow scrambler. Not in the Highlands, but found in Ireland, whence the name.

Menyanthes trifoliata—Bog-bean, buck-bean, marsh trefoil. Gaelic and Irish : *pònair chapull*, the mare's bean. (See *Faba*.) *Pacharan chapull*, the mare's packs or wallets, from *pac*, a pack, a wallet, a bundle. *Tri-bhileach*, the three-leaved plant. *Mill-sean monaidh*, the sweet plant of the hill.

> "*Millseineach*, biolaireach sobhrách."—M'LACHUINN.
> Abounding in bog-beans, cresses, primroses.

" The Highlanders esteem an infusion or tea of the leaves as good to strengthen a weak stomach" (Stuart).

CONVOLVULACEÆ.

Convolvulus arvensis—Field bindweed. Gaelic : *iadh lus*, the plant that surrounds. (See *Hedera helix*.)

C. sepium—Great bindweed. Gaelic and Irish : *dùil mhial* (Shaw), from *dùl*, catch with a loop ; and *mial*, a louse,—really signifying the plant that creeps and holds by twining.

49

Cuscuta epilinum—Flax dodder. Irish: *clamhainin lin*, the flax kites. It is parasitical on flax, to the crops of which it is sometimes very destructive. *Cunach* or (Gaelic) *conach*, that which covers, as a shirt, a disease. A general name applicable to all the species. Welsh: *llindag*, the flax choker.

SOLANACEÆ.

Solanum dulcamara—Bitter-sweet; woody nightshade. Gaelic and Irish: *searbhag mhilis*, bitter-sweet (Highland Society's Dictionary). *Fuath gorm*, the blue demon (*fuath*, hate, aversion, a demon). *Miotag bhuidhe*. Irish: *miathog buidhe*, the yellow nipper, pincher, or biter. *Slat ghorm* (*slat*, a wand, a switch; *gorm*, blue).

S. tuberosum—Potato. Gaelic: *bun-tata*, adaptation of the Spanish *batata*. Sir John M'Gregor has ingeniously rendered the word *bun-taghta*, a choice root!

Atropa belladona — Deadly nightshade; dwale banewort. Gaelic and Irish: *lus na h'oidhche*, the nightweed, on account of its large black berries and its somniferous qualities. Buchanan relates the destruction of the army of Sweno, the Dane, when he invaded Scotland, by the berries of this plant, which were mixed with the drink with which, by their truce, they were to supply the Danes, which so intoxicated them that the Scots killed the greater part of the Danish army while they were asleep. Welsh: *y gysiadur*, the putter to sleep.

Hyoscyamus niger — Henbane. Gaelic and Irish: *gafann* (*gabhann*), the dangerous one. *Detheogha, deodha, deo*, breath, that which is destructive to life. *Caoch-nan-cearc*, that which blinds the hens. Its seeds are exceedingly obnoxious to poultry, hence the English name henbane. The whole plant is a dangerous narcotic. Welsh: *slewyg yr iâr*, preventing or curing faintness.

SCROPHULARIACEÆ.

Verbascum thapsus—Mullein; hag's taper; cow's lungwort. Gaelic and Irish: *cuineal Mhuire*, or *cuingeal Mhuire* (from *cuing*, asthma, or shortness of breath. In pulmonary diseases of cattle it is found to be of great use, hence the name, cow's lungwort, or *cuinge*, narrowness, straightness, from its high, tapering stem; *Mhuire*, Mary's).

Veronica beccabunga—Brooklime. Gaelic: *lochal*, from *loch*, a lake, a pool, the pool-weed or lake-weed, being a water-plant. *Lothal* (*lo*, water). Irish: *biolar mhùin*, the contemptible cress; *mùn*, urine. Welsh: *llychlys y dwfr*, squatter in the water.

G

V. officinale—Common speedwell. Gaelic and Irish : *lus cré,* the dust-weed. *Seamar chre* (see *Oxalis*).

V. anagallis—Water-speedwell. Irish : *fualachtar, fual,* water, the one that grows in the water.

Euphrasia officinalis—Eyebright. Gaelic : *lus nan leac,* the hillside plant ; *leac,* a declivity. *Soillseachd nan sùil, soillse na sùl* (M'Donald), that which brightens the eye. *Rein an ruisg* (Stuart), water for the eye. *Glan ruis,* the eye-cleaner. Lightfoot mentions that the Highlanders of Scotland make an infusion of it in milk, and anoint the patient's eyes with a feather dipped in it, as a cure for sore eyes. Irish : *radharcain (radhairc),* sense of sight. *Lin radharc (lin,* the eye, wet), the eye wetter or washer. *Raeimin-radhairc (reim,* power, authority), that which has power over the sight. *Roisnin, rosg,* the eye, eyesight. *Caoimin (caoimh),* clean. Welsh : *gloywlys,* the bright plant. *'Llysieuyn eufras,* the herb Euphrasia (from ευφραινω, *euphraino,* to delight, from the supposition of the plant curing blindness). Arnoldus de Villa saith, " It has restored sight to them that have been blind a long time before ; and if it were but as much used as it is neglected, it would half spoil the spectacle trade " (Culpepper).

Pedicularis sylvatica—Dwarf red rattle. Irish : *lusan grolla.*

P. palustris—Louse-wort ; red rattle. Gaelic : *lus riabhach,* the brindled plant, possibly a contraction of *riabhdheargach* (Irish), red-streaked, a name which well describes the appearance of the plant. *Modhalan dearg,* the red modest one. *Lus na mial,* louse-wort, from the supposition that sheep that feed upon it become covered with vermin. *Bainne ghabhar,* goat's milk, from the idea that when goats feed on it they yield more milk. Its beautiful pink flowers were used as a cosmetic.

> " Sail-chuach 's *bainne ghabhar,* ⋅
> Suadh ri t' aghaidh,
> 'S cha n' neil mac righ air an domhain,
> Nach bi air do dheidh."

> Rub thy face with violet and goat's milk,
> And there is no prince in the world
> Who will not follow thee.

Rhinanthus crista-galli—The yellow rattle. Gaelic : *modhalan bhuidhe,* the yellow modest one. *Bodach na claiginn.* Irish : *bodan na cloigin,* the old man with the skulls. *Claigeann* or (Irish) *cloigoin,* a skull, from the skull-like appearance of its inflated calyces.

Scrophularia nodosa—Figwort. Gaelic : *lus nan cnapan,* the knobbed plant, from its knobbed roots. Old English: kernel-wort. *Donn-lus*, brown-wort, from the brown tinge of the leaves. *Farach dubh* (*faracha*, Irish), a beetle or mallet; *dubh*, dark. Wasps and beetles resort greatly to its small mallet-like flowers. Irish: *fotrum* (*fot, fothach*), glandered—from the resemblance of its roots to tumours. In consequence of this resemblance it was esteemed a remedy for all scrofulous diseases; hence the generic name *Scrophularia.*

Digitalis purpurea—Foxglove. Gaelic: *lus-nam-ban-sìth*, the fairy women's plant. *Meuran sìth* (Stuart), the fairy thimble. Irish: *an siothan* (*sioth*, Gaelic: *sìth*) means peace. *Sìthich*, a fairy, the most active sprite in Highland and Irish mythology. *Meuran*[1] *nan daoine marbh*, dead men's thimbles. *Meuran nan caillich mharbha*, dead women's thimbles. In Skye it is called *ciochan nan cailleachan marblia* (Nicolson), the dead old women's paps. Irish: *sian sleibhe*. (*Sian*, a charm or spell, a wise one, a fox; *sleibhe*, a hill). Welsh: *mcnyg ellyllon*, fairy's glove. O'Reilly gives another Irish name, *bolgan beic* (diminutive of *bolg*, a sack, a bag. Greek, Βολγος, *beic*, bobbing, curtseying). And frequently in the Highlands the plant is known by the familiar name, *an lus mòr*, the big plant. *Lus a bhalgair* (Aberfeldy), the fox-weed.

OROBANCHACEÆ.

(From Greek, ορoβoς, *orobos*, a vetch, and αγχειν, to strangle, in allusion to the effect of these parasites in smothering and destroying the plants on which they grow.) The name *mùchog* (from *mùch* smother, extinguish, suffocate) is applied to all the species.

O. major and minor—Broom-rape. Irish and Gaelic: *siorra-lach*, (Shaw)—*sior*, vetches, being frequently parasitical on leguminous plants; or' *siorrachd*, rape.

VERBENACEÆ.

Verbena officinalis—Vervain. Gaelic and Irish: *trombhod,*—*trom*, a corruption of *drum*, from Sanscrit *dru*, wood; hence Latin, *drus*, an oak, and *bod* or *bòid*, a vow. Welsh: *dderwen fendigaid*, literally, blessed oak,—the "herba sacra" of the ancients. Vervain was employed in the religious ceremonies of the Druids. Vows were made and treaties were ratified by its means. "Afterwards all sacred evergreens, and aromatic herbs, such as holly,

[1] *Meuran* and *digitalis* (*digitabulum*), a thimble, in allusion to the form of the flower.

rosemary, &c., used to adorn the altars, were included under the term verbena" (Brockie). This will account for the name *trombhod* being given by O'Reilly as "vervain mallow;" M'Kenzie, "ladies' mantle;" and Armstrong, "vervain."

LABIATÆ.

(From Latin, *labium*, a lip, plants with lipped corollæ.) Gaelic: *lusan lipeach*, or *bileach.*

Mentha—(From Greek Mινθη, *minthe*. A nymph of that name who was changed into mint by Prosperine, in a fit of jealousy, from whom the Gaelic name *mionnt* has been derived.) Welsh: *myntys.*

Mentha sylvestris—Horse-mint. Gaelic: *mionnt each*, horse-mint; *mionnt fiadhain*, wild mint; and if growing in woods, *mionnt choille*, wood-mint.

M. arvensis—Corn-mint. Gaelic: *mionnt an arbhair*, corn-mint.

M. aquatica—Water-mint. Gaelic: *cairteal.* Irish: *cartal, cartloin*, probably meaning the water-purifier, from the verb *cartam*, to cleanse, and *loin*, a rivulet, or *lon*, a marsh or swampy ground. *Misimean dearg* (Armstrong), the rough red mint. The whole plant has a reddish appearance when young.

M. viridis—Garden-mint, spear-mint. Gaelic: *mionnt ghàraidh*, the same meaning; and *meanntas*, another form of the same name, but not commonly used.

"Oir a ta sibh a toirt an deachaimh as a mhionnt."—STUART.

For ye take tithe of mint.

M. pulegium—Pennyroyal. Gaelic: *peighinn rioghail*, the same meaning.

"Am bearnan bride 's a *pheighinn-rioghail*."—M'INTYRE.

The dandelion and the *pennyroyal.*

Welsh: *coluddlys*, herb good for the bowels. *Dail y gwaed*, blood leaf.

Calamintha—Basil-thyme, calamint. Gaelic: *calameilt* (from Greek, καλος, beautiful; and μινθη, *minthe*, mint), beautiful mint. Welsh: *Llysie y gâth*, cat-wort.

Rosmarinus officinalis—Common rosemary. Gaelic: *ròs Mhuire.* Irish: *ròs-mar—mar-ros*, sea-dew, corruptions from the Latin (*ros*, dew, and *marinus*), the sea-dew. *Ròs Mhairi*, Mary's rose, or rosemary. Welsh: *ròs Mair.* Among Celtic tribes rosemary was the symbol of fidelity with lovers. It was frequently worn at weddings. In Wales it is still distributed among friends at funerals, who throw the sprigs into the grave over the coffin.

Lavendula spica — Common lavender. Gaelic : *lus-na-tùise*, the incense plant, on account of its fragrant odour. *An lus liath*, the grey weed. *Lothail*, " *uisge an lothail*," lavender-water.

Satureia hortensis—Garden savory. Gaelic : *garbhag ghàraidh*, the coarse or rough garden plant, from *garbh*, rough, &c.

Salvia verbenacea—Clary. The Gaelic and Irish name, *torman*, applies to the genus as well as to this plant; it simply means " the shrubby one " (*tor*, a bush or shrub). The genus consists of herbs or undershrubs, which have generally a rugose appearance. A mucilage was produced from the seeds of this plant, which, applied to the eye, had the reputation of clearing it of dust; hence the English name, " clear-eye," clary (Gaelic : *clearc*, bright).

S. officinalis—Garden-sage (of which there are many varieties). Gaelic : *athair liath*, the grey father. *Sàisde* (from *sage*). *Slàn lus*, the healing plant, corresponding with *salvia* (Latin : *salvere*, to save). It was formerly of great repute in medicine. Armstrong remarks : " Bha barail ro mhòr aig na sean Eadalltich do 'n lus so, mar a chithear o'n rann a leanas,—

> " Cur moriatur homo cui *salvia* crescit in horto ? "
> C' arson a gheibheadh duine bàs,
> Aig am bheil *sàisde* fàs na gharaidh ?
> Why should the man die who has *sage* growing in his garden ?

Teucrium scorodonia — Wood - sage. Gaelic : *sàisde coille*, wood-sage. *Sàisde fiadhain*, wild sage. O'Reilly gives the name *ebeirsluaigh*, perhaps from *obar*, shall be refused, and *sluagh*, people, multitude, because it did not possess the virtues attributed to the other species, and even cattle refuse to eat it.

Thymus serpyllum—Thyme, wild thyme. Gaelic and Irish : *lus mhic righ Bhreatainn*, the plant belonging to the king of Britain's son. *Lus an righ*, the king's plant. This plant had the reputation of giving courage and strength through its smell ; hence the English thyme (from Greek : θυμος, *thymos*, courage, strength,—virtues which were essential to kings and princes in olden times). Highlanders take an infusion of it to prevent disagreeable dreams. Welsh : *teim*.

Origanum { **marjorana** / **vulgare** } —Marjoram. Gaelic and Irish : *oragan*, the delight of the mountain. Greek : ορος, *oros*. Gaelic : *ord*, a mountain ; and Greek, γανος, *ganos*, joy. Gaelic : *gain*, clapping of hands. *Lus mharsalaidh*, the merchant's weed, may only be a corrupted form of marjoram, from an Arabic word (*maryamych*). *Seathbhog*, the skin or hide softener (*seathadh*, a skin, a hide, and *bog*, soft). "The dried leaves are used in fomentations, the essential oil is so acrid that it may be considered as a caustic, and was formerly used as such by farriers" (Don). Welsh: *y benrudd*, ruddy-headed.

O. dictamnus—Dittany. The Gaelic and Irish name, *lus a phiobaire*—given in the dictionaries for " dittany "—is simply a corruption of *lus a pheubair*, the pepperwort, and was in all probability applied to varieties of *Lepidium* as well as to *Origanum dictamni creti*, whose fabulous qualities are described in Virgil's 12th ' Æneid,' and in Cicero's ' De Natura Deorum.'

Hyssopus officinalis— Common hyssop. Gaelic: *isop*. French: *hysope*. German : *isop*. Italian : *isopo* (from the Hebrew name, אֵזוֹב, *ezob*, or Arabian, *azzof*).

> "Glan mi le *h' isop*, agus bithidh me glan."
> Purge me with *hyssop*, and I shall be clean.

Ajuga reptans—Bugle. Gaelic : *meacan dubh fiadhain* (Armstrong), the dusky wild plant. Welsh : *glesyn y coed*, wood-blue.

Nepeta glechoma— Ground-ivy. Gaelic : *iadh shlat thalmhainn*, the ground-ivy. (See *Hedera helix*, and *Bunium flexuosum*). *Nathair lus*, the serpent-weed,—it being supposed to be efficacious against the bites of serpents ; hence the generic name, *Nepeta*, from *nepa*, a scorpion. Irish: *aignean thalmhuin* (*aigne*, affection, *thalmhuin*, the ground) ; *eidhnean thalmhuin* (see *Hedera helix*).

Ballota niger—Stinking horehound. Irish and Gaelic : *gràfan* or *gràbhan dubh*, the dark opposer (*grab*, to hinder or obstruct). It was a favourite medicine for obstructions of the viscera : or it may refer to *grab*, a notch, from its indented leaves.

Lycopus europæus— Water-horehound. Irish : *feoran curraidh*, the green marsh-plant (*currach*, a marsh).

Marrubium vulgare—White horehound. Gaelic and Irish : *gràfan* or *gràbhan bàn*, the white indented, &c. (See *Ballota niger*).

Lamium album — White dead-nettle ; archangel. Gaelic: *teanga mhìn*, the smooth tongue. *Ionntag bhàn*, white nettle. *Ionntag mhàrbh*, dead nettle. (For *Ionntàg* see *Urtica*.)

L. purpureum — The red dead-nettle. Gaelic : *ionntag dhearg*, red nettle.

Galeopsis tetrahit—Common hemp-nettle. Gaelic: *an gath dubh*, the dark bristly plant (*gath*, a sting, a dart). It becomes black when dry, and has black seeds.

G. versicolor—Large-flowered hemp-nettle. Gaelic :. *an gath buidhe,—an gath mòr*, the yellow bristly plant—the large bristly plant. Very abundant in the Highlands, and troublesome to the reapers at harvest-time, from its bristly character. It is called yellow on account of its large yellow flower, with a purple spot on the lower lip.

Stachys betonica—Wood-betony. Gaelic: *lus bheathag*, the life-plant, nourishing plant (from Irish: *beatha;* Greek: βιωτα; Latin: *vita,*—life, food). "*Betonic*, a Celtic word; *ben*, head, and *ton*, good, or tonic" (Sir W. J. Hooker). *Biatas* (from *biadh*, feed, nourish, maintain). "A precious herb, comfortable both in meat and medicine" (Culpepper). *Glasair coille*, the wood green one. The green leaves were used as a salad: any kind of salad was called *glasag*.

S. sylvatica—Wound-wort. Gaelic: *lus nan scorr*, the wound-wort (*scorr*, a cut made by a knife or any sharp instrument). Irish: *caubsadan*.

Prunella vulgaris—Self-heal. Gaelic and Irish: *dubhan ceann chòsach*, also *dubhanuith*. These names had probably reference to its effects as a healing plant. "It removes all obstructions of the liver, spleen, and kidneys" (*dubhan*, a kidney, darkness; *ceann*, head, and *còsach*, spongy or porous). *Slàn lus*, healing plant. *Lus a chridh*, the heart-weed. Irish: *ceanabhan-beg*, the little fond dame; *cean*, fond, elegant, and *ban*, woman, wife, dame.

BORAGINACEÆ.

Borago officinalis—Borage. Gaelic and Irish: *borrach, borraist, borraigh*, all these forms are evidently derived from *borago*, altered from the Latin, *cor*, the heart, and *ago*, to act or effect. The plant was supposed to give courage, and to strengthen the action of the heart; "it was one of the four great cordials." *Borr* in Gaelic means to bully or swagger; and *borrach*, a haughty man, a man of courage. Welsh: *llawenllys* (*llawen*, merry, joyful), the joyful or glad plant.

Lycopsis arvensis—Bugloss. Gaelic: *lus-teang' an daimh*, ox-tongue. *Boglus*, corruption of *bolg*, an ox; *lus*, a plant. Welsh: *tafod yr ych*, the same meaning. *Bugloss*, from Greek βους, *bous*, an ox, and γλωσσα, *glossa*, a tongue, in reference to the roughness and shape of the leaves.

Myosotis palustris—Marsh scorpion-grass or forget-me-not. Gaelic and Irish : *cotharach*, the protector (*cothadh*, protection) ; perhaps the form of the racemes of flowers, which, when young, bend over the plant as if protecting it. *Lus nam mial*, the louse-plant,—probably a corruption from *miagh*, esteem. *Lus midhe* (O'Reilly), a sentimental plant that has always been held in high esteem.

Symphytum officinale—Comfrey. Gaelic : *meacan dubh*, the large or dark plant. Irish : *lus na cnamh briste*, the plant for broken bones. The root of comfrey abounds in mucilage, and was considered an excellent remedy for uniting broken bones. "Yea, it is said to be so powerful to consolidate and knit together, that if they be boiled with dissevered pieces of flesh in a pot, it will join them together again " (Culpepper).

Echium vulgare—Viper's bugloss. *Boglus* (see *Lycopsis*) and *us na nathrach*, the viper's plant.

Cynoglossum officinale — Common hound's-tongue. Gaelic and Irish : *teanga con* (O'Reilly). *Teanga chù*, dog's-tongue. Welsh : *tafod y ci*, same meaning. Greek : *cynoglossum* (κυων, *kyon*, a dog, and γλωσσα, *glossa*, a tongue), name suggested from the form of the leaves.

PINGUICULACEÆ.

Pinguicula vulgaris—Bog-violet. Gaelic : *bròg na cubhaig*, the cuckoo's shoe, from its violet-like flower. *Badan measgan*, the butter mixer ; *badan*, a little tuft, and *measgan*, a little butter-dish ; or *measg*, to mix, to stir about. On cows' milk it acts like rennet. *Lus a bhainne*, the milk-wort. It is believed it gives consistence to milk by straining it through the leaves. *Uachdar*, surface, top, cream, — a name given because it was supposed to thicken the cream.

PRIMULACEÆ.

Primula vulgaris—Primrose. Gaelic : *sobhrach, sobhrag*.

> " A *shobhrach*, geal-bhui nam bruachag,
> Gur fan-gheal, snughar, do ghnùis !
> Chinneas badanach, cluasach,
> Maoth-mhìn, baganta luaineach.
> Bi'dh tu t-eideadh sa'n earrach
> 'S 'càch ri falach an sùl."—M'DONALD.

> Pale yellow primrose of the bank,
> So pure and beautiful thine appearance !
> Growing in clumps, round-leaved,

Tender, soft, clustered, waving ;
Thou wilt be dressed in the spring
When the rest are hiding in the bud.

The Irish name *soghradhach* (Shaw), means amiable, lovely, acceptable. The Gaelic names have the same meaning. *Sobh* or *subh*, pleasure, delight, joy. *Soradh, soirigh*, are contractions ; also *samharcan*. Irish : *samharcain* (*samhas*, delight, pleasure).

> "Am bi na *samhraichean* s' neoinean fann."—OLD SONG.
> "Gu tric anns' na bhuinn sinn a t' *sòrach*."—MUNRO.
> Often we gathered there the *primrose*.

Welsh : *briollu*,—*briol*, dignified ; *allwedd*, key. "The queenly flower that opens the lock to let in summer " (Brockie).

P. veris — Cowslip. Gaelic : *muisean*, the low rascal, the devil. "*A choire mhuiseanaich*," a dell full of cowslips. Cattle refuse to eat it, therefore farmers dislike it. *Bròg na cubhaig* (M'Kenzie), the cuckoo's shoe. Irish : *seichearlan, seicheirghin, seicheirghlan*, from *seiche*, hide or skin. It was formerly boiled, and "an ointment or distilled water was made from it, which addeth much to beauty, and taketh away spots and wrinkles of the skin, sun-burnings and freckles, and adds beauty exceedingly." The name means the "skin-purifier." *Baine bó bhuie*, the yellow cow's milk. *Baine bo bleacht*, the milk-cow's milk.

P. auricula—Auricula. Gaelic : *lus na bann-righ*, the queen's flower.

P. Polyanthus—Winter primrose. Gaelic : *Sobhrach gheamh-raidh*.

Cyclamen hederæfolia—Sow-bread. Gaelic : *culurin* (perhaps from *cul* or *cullach*, a boar, and *aran*, bread), the boar's bread.

Lysimachia (from Greek λυσω and μαχὅμαι, I fight).

L. vulgaris—Loose-strife. Gaelic and Irish : *lus na sìthchaine*, the herb of peace (*sìth*, peace, rest, ease ; *cáin*, state of). *Conaire*, the keeper of friendship. The termination "*aire*" denotes an agent ; and *conall*, friendship, love. *An seileachan buidhe*, the yellow willow herb.

L. nemorum—Wood loose-strife ; yellow pimpernel. Gaelic and Irish : *seamhair Mhuire* (*seamhair, seamh*, gentle, sweet, and *feur*, grass ; *seamhrog* (shamrock), generally applied to the trefoils and wood-sorrel. (See *Oxalis*.) *Mhuire* of Mary ; *Maire*, Mary. This form is especially applied to the Blessed Virgin Mary. In the Mid-Highlands more frequently called *Samman* (Stewart). *Lus Cholum-cille*, the wort of St Columba, the apostle

H

58

of Scotland. *Columb*, a dove; *cillé*, of the church. This name is given in the Highlands to *Hypericum*, which see. *Rosor* (O'Reilly). *Ros* is sometimes used for *lus*. *Ros-or*, yellow or golden rose. "From the Sanskrit, *ruksha* or *rusha*, meaning tree, becomes in Gaelic *ros*, a tree or treelet, just as *daksha*, the right hand, becomes *dexter* in Latin and *deas* in Gaelic. *Ros*, therefore, means a tree or small tree, or a place where such trees grow—hence the names of places that are marshy or enclosed by rivers, as Roslin, Ross-shire, Roscommon," &c.—CANON BOURKE.

Anagallis arvensis — Pimpernel, poor man's weather-glass. Gaelic : *falcair*. Irish : *falcaire fiodhain*, the wild cleanser (*falcadh*, to cleanse). The name expressing the medicinal qualities of the plant, which, by its purgative and cleansing power, removes obstructions of the liver, kidneys, &c. *Falcaire fuair*, —*falcaire* also means a reaper, and *fuair*, cold ; *fuaradh*, to cool, a weather-gauge. The reaper's weather-gauge, because it points out the decrease of temperature by its hygrometrical properties —when there is moisture the flower does not open. *Loisgean* (M'Donald), from *loisg*, to put in flame, on account of its fiery appearance. *Ruinn ruise* (O'Reilly). *Ruinn* means sex, and by pre-eminence the "male ;" *ruise* is the genitive case of *ros*. It is still called the male pimpernel in some places. The distilled water or juice of this plant was much esteemed formerly for cleansing the skin.

PLUMBAGINACEÆ.

Armeria maritima—Thrift. Gaelic : *tonn a chladaich* (Armstrong), the "beach-wave," frequent on the sea-shore, banks of rivers, and even on the Grampian tops. *Bàr-dearg*, red-top, from its pink flower. *Nebinean chladaich*, the beach daisy, from *clàdach*, shore, beach, sandy plain.

PLANTAGINACEÆ.

Plantago major—Greater plaintain. Gaelic and Irish : *cuach Phàdraig*, Patrick's bowl or cup,—in some places *cruach Phàdraig*, Patrick's heap or hill. Welsh : *llydain y fford*, spread on the way.

P. lanceolata—Rib-wort. Gaelic and Irish : *slàn lus*, the healing plant.

" Le meilbheig, le nebinean 's le *slàn-lus*."—M'LEOD.
With poppy, daisy, and *rib-wort*.

Lus an t' slanuchaidh (*lus*, a wort, a plant-herb, chiefly used for plant; it signifies also power, force, efficacy; *slanuchaidh*, a participial noun from *slan*; Latin, *sanus*), the herb of the healing, or healing power; a famous healing plant in olden times. *Deideag.* Irish: *deideog* (*ag* and *òg*, young, diminutive terminations; *deid*, literally *deud* or *deid*, a tooth), applied to the row of teeth, and also to the nipple (Gaelic, *diddi*; English, *titty*), because like a tooth, hence to a plaything,—play, *gewgaw*, bo-peep, a common word with nurses.

> " B'iad sid an geiltre glé ghrinn.
> Cinn *deideagan* measg feoir," &c.—M'DONALD.
>
> Scenes of startling beauty,
> Plantain-heads among the grass, &c.

Armstrong translates it "gewgaws" amongst the grass; but the editor of ' Sar-obair nam Bard Gaelach '—see his vocabulary— gives *deideagan*, rib-grass, which renders the line intelligible. *Bodaich dhubha*, the black men,—children's name in Perthshire. Welsh : *llwynhidydyl-penaùr*.

PARONYCHIACEÆ.

Herniaria glabra—Rupture-wort; burst - wort. Gaelic and Irish : *lus an t' sicnich* (M'Kenzie), from *sic*, the inner skin that is next the viscera in animals. "*Bhrist an t sic*," the inner skin broke. "*Mam-sic*," rupture, hernia. Not growing naturally in Scotland, but was formerly cultivated by herbalists as a cure for hernia.

CHENOPODIACEÆ.

Amaranthus caudatus — Love-lies-bleeding. Gaelic: *lus a ghràidh*, the love plant. *Gràdh*, love.

Spinacia oleracea — Spinage. Gaelic : *bloinigean gàraidh*. *Blonag*, fat (Welsh, *bloneg*; Irish, *blanag*); *gàradh*, a garden. *Slàp chàil* (M'Alpin); *slàp*, to flap; *càl*, cabbage. Welsh : *yspigoglys*.

Beta maritima—Beet, mangel-wurzel. Gaelic : *betis, biotas*. Irish : *biatas*. Welsh : *beatws* (evidently on account of its feeding or life-giving qualities). Greek : βίος. Latin : *vita*, life, food ; and the Gaelic : *biadh*, feed, nourish, fatten. Cornish : *boet*.

Suæda maritima—Sea-side goose grass. } Gaelic and Irish :
Salicornia herbacea—Glass-wort. } *praiseach na màra*, the sea pot-herb. Name applied to both plants. For *praiseach*, see *Crambe maritima*.

Atriplex hastata and **patula**—Common orache. Gaelic and Irish: *praiseach mhìn*. *Mìn*, meal, ground fine, small. Still used by poor people as a pot-herb. *Ceathramha-luain-griollog* (O'Reilly), loin-quarters. *Ceathramadh caorach* (Bourke), sheep's quarters. The name *griollog* is applied also to the samphire.

A. portulacoides — Purslane-like orache. Gaelic and Irish: *purpaidh*, purple. A name also given to the poppy. Name given on account of the purple appearance of the plant, it being streaked with red in the autumn.

Chenopodium vulvaria (or **olidum**) — Stinking goosefoot. Irish: *elefleog*. *El* or *ela*, a swan; and *flè* or *flèadh*, a feast. It was said to be the favourite food of swans. Scotch: *olour* (Latin, *olor*, a swan).

C. album — White goosefoot. Gaelic and Irish: *praiseach fiadhain*, wild pot-herb. The people of the Western Highlands, and poor people in Ireland, still eat it as greens. *Praiseach glàs*, green pot-herb, a name given to the fig-leaved goosefoot (*ficifolium*).

C. Bonus-Henricus—Good King Henry, wild spinage, English Mercury. Gaelic and Irish: *praiseach bràthair*, the friar's pot-herb. (*Bràthair* means brother, also friar—*frère*). Its leaves are still used as spinage or *spinach*, in defect of better.

LAURACEÆ.

Laurus (from Sanskrit *labhasa*, abundance of foliage; root *labh*, to take, to desire, to possess—akin to Greek, λαμβανω, *lambano*).—Gaelic: *lamh*, a hand (Canon Bourke).

L. nobilis—The laurel, the bay-tree (which must not be confounded with our common garden laurel, *Prunus lauro-cerasus* and *P. lusitanicus*). Gaelic and Irish: *labhras*. *Crann laoibh-reil*, the tree possessing richness of foliage. With its leaves poets and victorious generals were decorated. The symbol of triumph and victory. It became also the symbol of massacre and slaughter, hence another Gaelic name, *casgair*, to slaughter, to hit right and left. *Ur uaine*, the green bay-tree.

"Agus e' ga sgaoileadh fèin a mach mar *ùr chraoibh uaine.*"
And spreading himself like a green *bay-tree.*—PSALM xxxvii. 35.

Ur = bay or palm tree, from the Sanskrit, *ùrh*, to grow up. Palm Sunday is styled "*Domhnach an ùir*," the Lord's day of the palm.

L. cinnamomum—Cinnamon. Gaelic and Irish: *caineal*.

6

" 'Se 's millse ua 'n *caineal.*"—BEINN-DORAIN.
It is sweeter than *cinnamon.*

Canal (Welsh : *canel*).

" Rinn mi mo leabadh cùbhraidh le mirr, aloe, agus *canal.*"—PROVERBS
vii. 17.
I have perfumed my bed with myrrh, aloes, and cinnamon.

From the Hebrew : קִנָּמוֹן, *qinnamon.* Greek : κινάμωμον, *kina-
mōmon.*

POLYGONACEÆ.

Polygonum (from πολυς, many, and γονυ, knee, many knees
or joints).—Gaelic : *lusan gluineach,* kneed or jointed plants. ˜
Polygonum bistorta—Bistort, snakeweed. Gaelic and Irish :
bilur (O'Reilly). Seems to mean the same as *biolair,* a water-
cress. The young shoots were formerly eaten. Welsh : *lysiau'r
neidr,* adder's plant.

P. amphibium—Amphibious persicaria. Gaelic and Irish :
gluineach an uisge, the water-kneed plant. It is often floating
in water. *Gluineach dhearg,* the red-kneed plant. Its spikes of
flowers are rose-coloured and handsome. Armstrong gives this
name to *P. convolvulus,* which is evidently wrong.

P. aviculare—Knot-grass. Gaelic and Irish : *gluineach bheag*
(O'Reilly), the small-jointed plant.

P. convolvulus — Climbing persicaria ; black bindweed ;
climbing buckwheat. Gaelic and Irish : *gluineach dhubh,* the
dark-jointed plant.

P. persicaria—The spotted persicaria. Gaelic and Irish :
gluineach mhòr, the large-jointed plant. *Am boinne-fola* (Fergusson), the blood-spot. *Lus chrann ceusaidh* (M'Lellan), herb of
the tree (of) crucifixion. The legend being that this plant grew
at the foot of the Cross, and drops of blood fell on the leaves,
and so they are to this day spotted.

P. hydropiper — Water-pepper. Gaelic : *lus an fhògair*
(M'Kenzie), the plant that drives, expels, or banishes. It had
the reputation of driving away pain, flies, &c. " If a good hand-
ful of the hot biting arssmart be put under the horse's saddle,
it will make him travel the better though he were half tired
before "—CULPEPPER. *Gluineach tèth,* the hot-kneed plant.

Rumex obtusifolius ⎫
   „   **crispus** ⎬—Dock. Gaelic and Irish : *còpag*—
   „   **conglomeratus** ⎭
còpagach, còpach, bossy. Welsh : *copa,* tuft, a top.

R. sanguineus—Bloody-veined dock. Gaelic: *a chòpagach dhèarg*, the red dock. The stem and veins of leaves are blood-red.

R. alpinus—Monk's rhubarb. Gaelic : *lus na purgaid*, the purgative weed. A naturalised plant. The roots were formerly used medicinally, and the leaves as a pot-herb. Welsh : *arian-llys*. The same name is given for rue.

R. acetosa—Common sorrel. Gaelic : *samh*, sorrel. Irish : *samhadh bo*, cow-sorrel (for *samh* see *Oxalis*). *Puinneag* (M'Donald). Irish : *puineoga*. Name given possibly for its efficacy in healing sores and bruises (a pugilist, *puinneanach*). *Sealbhag*, not from *sealbh*, possession, more likely from *searbh*, sour, bitter, from its acid taste.

> " Do *shealbhag* ghlàn 's do luachair
> A bòrcadh suas ma d' choir."—M'Donald.
>
> Thy pure *sorrel* and thy rushes
> Springing up beside thee.

Sealgag (Irish, *sealgan*), are other forms of the same name. *Copog shraide*, the roadside or lane dock. *Sobh* (Shaw), the herb sorrel.

R. acetosella—Sheep's sorrel. Gaelic and Irish : *ruanaidh*, the reddish-coloured. It is often bright red in autumn. *Pluirin seangan* (O'Reilly), the small-flowered plant (*pluran*, a small flower ; *seangan*, slender). *Samhadh caora* (O'Reilly), sheep's sorrel.

Oxyria reniformis — Mountain-sorrel. Gaelic and Irish : *sealbhaig nan fiadh*, the deer's sorrel.

ARISTOLOCHIACEÆ.

Aristolochia clematitis—Birth-wort. *Culurin* (see *Cyclamen*).

Asarum europæum — Common asarum. Gaelic : *asair* (M'Donald), from the generic name, said to be derived from Greek—a, privative, and σειρα, bandage. The leaves are emetic, cathartic, and diuretic. The plant was formerly employed to correct the effects of excessive drinking, hence the French, *cabaret*.

EMPETRACEÆ.

Empetrum nigrum—Crow-berry. Gaelic and Irish : *lus na fionnag* (*fionnag*, a crow). Sometimes written *fiannag*, *fiadhag* (*dearc fithich*, raven's berry ; *caor fionnaig*, crow-berry), the ber-

ries which the Highland children are very fond of eating, though
rather bitter. Taken in large quantities, they cause headache.
Grouse are fond of them. Boiled with alum they are used to
produce a dark-purple dye. *Lus na stalog* (O'Reilly), the star-
ling's plant.

EUPHORBIACEÆ.

Euphorbia exigua
    „   **helioscopia** } —Spurge. Gaelic and Irish : *spuirse*
= spurge. *Foinneamh lus*, wart-wort.

E. peplus—Petty spurge. Gaelic and Irish : *lus leusaidh*,
healing plant. The plants of this genus possess powerful cathartic and emetic properties. *E. helioscopia* has a particularly
acrid juice, which is often applied for destroying warts, hence it
is called *foinneamh lus*. Irish : *gear neimh* (*gear* or *geur*, severe,
and *neimh*, poison, the milky juice being poisonous.)

E. paralias—Sea-spurge. Irish : *buidhe na ningean* (O'Reilly),
the yellow plant of the waves (*nin*, a wave), its habitat being
maritime sands. Not found in Scotland, but in Ireland, on the
coast as far north as Dublin. This and the preceding species
are extensively used by the peasantry of Kerry for poisoning, or
rather stupefying, fish.

Buxus sempervirens — Box. Gaelic and Irish : *bocsa*, an
alteration of βυξος, the Greek name.

"Suidhichidh mi anns an fhàsach an giuthas, an gall ghiùthas, agus am
bocsa le cheile."—ISAIAH.

I will set in the desert the fir-tree and the pine and the *box* together.

The badge of Clan M'Pherson and Clan M'Intosh.

Mercurialis perennis—Wood mercury. Gaelic : *lus ghlinne-
bhracadail*. *Lus ghlinne*, the cleansing wort ; *bracadh*, suppura-
tion, corruption, &c. It was formerly much used for the cure
of wounds.

CUCURBITACEÆ.

Cucumis sativus—Cucumber. Gaelic and Irish : *cularan*,
perhaps from *culair*, the palate, or *culear*, a bag.

"Is cuimhne leinne an t-iasg a dh 'ith sinn san Ephit gu saor ; na-*cular-
ain* agus na *mealbhucain*."—NUMBERS xi. 5.

We remember the fish that we did eat in Egypt freely, and the cucumber
and the melons.

"'Sa thorc nimhe ri sgath a *chularan*."—M'DONALD.

The wild boar destroying his *cucumbers*.

Irish : *cucumhar* (O'Reilly), cucumber, said to be derived from the Celtic word *cuc* (Gaelic, *cuach*), a hollow thing. In some species the rind becomes hard when dried, and is used as a cup. Latin : *cucurbita*, a derivative from the Celtic. (See Loudon.) Welsh : *chwerw ddwfr* = water-sour.

Cucumis melo—Melon. Gaelic and Irish : *meal-bhuc*, from *mel* or *mal* (Greek, μελον, an apple), and *buc*, size, bulk. According to Brockie, " *mealbhucain* (plural), round fruit covered with warts or pimples." *Mileog*, a small melon.

URTICACEÆ.

Urtica—A word formed from Latin : *uro*, to burn.

U. urens
 „ dioica } —Nettle (Anglo-Saxon, *nædl*, a needle). Gaelic and Irish : *feanntag, neandog,*[1] *deanntag, iontag, iuntag* (from *feannta,* flayed, pierced, pinched—*feann,* to flay, on account of its blistering effects on the skin ; *ang,* a sting ; *iongna,* nails). Latin : *ungues.*

" Sealbhaichidh an t' ionntagach."—HOSEA.
The nettles shall possess them.

To this day it is boiled in the Highlands and in Ireland by the country people in the spring-time. Till tea became the fashion, nettles were boiled in meal, and made capital food. *Caol-fàil*—*caol,* slender ; *fàl,* spite, malice. In the Hebrides often called *sradag* (a spark), from the sensation (like that from a fiery spark) consequent upon touching. (Stuart.)

Cannabis sativa—Hemp. Gaelic and Irish : *caineab,* the same as *cannabis,* and said to be originally derived from Celtic, *can,* white ; but the plant has been known to the Arabs from time immemorial under the name of *quaneb. Corcach,* hemp.

" Buill do' n chaol *chòreaidh.*" M'DONALD.
Tackling of hempen ropes.

Welsh : *cynarch.*

Parietaria officinalis — Wall pellitory. Gaelic and Irish : *lus a bhallaidh,* from *balladh* (Latin, *vallum* ; Irish, *balla*), a wall. A weed which is frequently found on or beside old walls or rubbish heaps, hence the generic name " parietaria," from

---

[1] "*Neandog,* the common name for it in Ireland. In feminine nouns, the first consonant (letter) after the article *an* (the) is softened in sound. 'An feanntag'—'f' when affected loses its sound, and 'N' is sounded instead : 'N (f)eantog.'"—CANON BOURKE.

paries, a wall. Irish: *mionntas chàisil* (*càisiol*, any stone building), the wall-mint. For *mionntas*, see *Mentha*.

Humulus lupulus—Hop. Gaelic and Irish: *lus an leanna— lionn luibh*, the ale or beer plant. *Lionn, leann* (Welsh, *lhyn*) beer, ale.

Ulmus—Elm. Celtic: *ailm*. The same in Anglo-Saxon, Teutonic, Gothic, and nearly all the Celtic dialects. Hebrew: אלה, *elah*, translated oak, terebinth, and elm.

U. campestris—Gaelic and Irish: *leamhan, slamhan* (Shaw), *liobhan*. Welsh: *llwyfen*. According to Pictet, in his work, 'Les Origines Indo-Europeennes ou les Aryas Primitifs,' p. 221, "To the Latin: 'Ulmus' the following bear an affinity (respond)—Sax.: *ellm*. Scand.: *almr*. Old German: *elm*. Rus.: *ilemu*. Polish: *ilma*. Irish: *ailm, uilm*, and by inversion, '*leamh*,' or '*leamhan*.'" He says the root is *ul*, meaning to burn. The tree is called from the finality of it, "to be burned." That is his opinion, and he is probably right. The common idea of *leamhan* is that it is from *leamh*, tasteless, insipid, from the taste of its inner bark; and *liobh* means smooth, slippery. And the tree in Gaelic poetry is associated with or symbolic of slipperiness of character, indecision. Cicely M'Donald, who lived in the reign of Charles II., describing her husband, wrote as follows:—

> " Bu tu' n t-iubhair as a choille,
>   Bu tu' n darach daingean làidir,
>   Bu tu' n cùileann, bu tu 'n droighionn,
>   Bu tu' n t' abhall molach, blàth-mhor,
>   Cha robh meur annad do' n chritheann,
>   Cha robh do dhlighe ri feàrna,
>   *Cha robh do chàirdeas ri leamhan,*
>   ' Bu tu leannan nam ban àluinn."

> Thou wast the yew from the wood,
>   Thou wast the firm strong oak,
> Thou wast the holly and the thorn,
>   Thou wast the rough, pleasant apple,
> Thou had'st not a twig of the aspen,
>   Under no obligation to the alder,
> *And hadst no friendship with the elm,*
>   Thou wast the beloved of the fair.

Ficus—Nearly the same in most of the European languages. Greek: συχη. Latin: *ficus*. Celtic: *fige*.

F. carica—Common fig-tree. Gaelic and Irish: *crann fìge* or *fìghis*.

I

"Ach fòghlumaibh cosamhlach do'n *chrann fhìge.*"
Learn a parable from the fig-tree.

Morus—Greek : μορος, *moros.* Latin : *morus,* a mulberry.
Loudon, in his 'Encyclopedia of Plants,' says it is from the
Celtic *mòr,* dark-coloured. There is no such Celtic root; it may
be from the Sanskrit, *murch,* Scotch, *mirk,* darkness, obscurity;
and the Greek name has also this meaning. The fruit being of
a darkish red colour. Old Ger. and Danish : *mur-ber.*

M. nigra — Common mulberry. Gaelic and Irish : *crànn-
maol-dhearc,* tree of the mild aspect, or if *dearc* here be a berry,
the mild-berry tree. *Maol* (Latin, *mollis*) has many significations. Bald, applied to monks without hair, as *Maol Cholum,*
St Columba ; *Maol Iosa, Maol Brighid,* St Bridget, &c. A pro-
montory, cape, or knoll, as *Maol Chinntìre,* Mull of Cantyre.
Malvern, *maol,* and *bearna,* a gap. To soften, by making it less
bitter, as "dean maol é," make it mild. Hence mulberry, mild-
berry (Canon Bourke).

AMENTIFERÆ AND CUPULIFERÆ.

Catkin-bearers—Gaelic : *caitean,* the blossom of osiers.

" 'Nis treigidh coileach á ghucag
'S *caitean* brucach nan craobh."—M'DONALD.

Now the cock will forsake the buds
And the spotted catkins of the trees.

Quercus—Said in botanical works to be from the Celtic, *quer,*
fine. There is no such word in any Celtic dialect, and even
Pictet has failed, after expending two pages on it, to explain it.

Q. robur—("Robur comes from the Celtic, *ro,* excelling, and
bur, development"—CANON BOURKE). The oak. Gaelic and
Irish : *dair,* genitive *dàrach,* sometimes written *dàrag, dùr, drù.*
Sanskrit : *dru, druma, druta,* a tree, the tree ; *daru,* a wood.

" Sàmhach' us mòr a bha 'n triath,
Mar dharaig 's i liath air Lùbar,
A chaill a dlu-dheug o shean
Le dealan glan nan spéur,
Tha 'h-aomadh thar srùth o shliabh,
A còinneach mar chiabh a fuaim."—OSSIAN.

Silent and great was the prince
Like an oak-tree hoary on Lubar,
Stripped of its thick and aged boughs
By the keen lightning of the sky,
It bends across the stream from the hill,
Its moss sounds in the wind like hair.

Om, omna, the oak (O'Reilly). "Cormac, King of Cashel, Ireland, A.D. 903, says of *omna* that it equals *fuamna,* sounds, or noises, because the winds resound when the branches of the oak resist its passage. According to Varro, it is from *os,* mouth, and *men,* mind, thinking—that is, telling out what one thinks is likely to come. Cicero agrees with this, 'Osmen voces hominum'"—CANON BOURKE. Compare-Latin : *omen,* a sign, a prognostication,—it being much used in the ceremonies of the Druids. *Omna,* a lance, or a spear, these implements being made from the wood of the oak. Greek : δopu, a spear, because made of wood or oak. *Eitheach,* oak, from *eithim,* to eat, an old form of *ith.* Latin : *ed-ere,* as "oak" is derived from *ak* (Old German) to eat (the acorn). The "oak" was called *Quercus esculus* by the Latins. *Rail, railaidh,* oak.

> "Ni bhiodh achd, aon dhearc ar an *ralaidh.*"
> There used to be only one acorn on the oak.

Canon Bourke thinks it is derived from *ro,* exceeding, and *ail,* growth ; or *ri,* a king, and *al* or *ail*—that is, king of the growing plants. The Highlanders still call it *righ na coille,* king of the wood. The Spanish name *roble* seems to be cognate with *robur.*

Q. ilex—Holm-tree. Gaelic and Irish : *craobh thuilm,* genitive of *tolm,* a knoll, may here be only an alteration of "holm." *Darach sior-uàine,* ever-green oak.

Q. suber—The cork-tree. Gaelic : *crànn àirceain.* Irish : *crann àirc. Airc,* a cork.

Fagus sylvatica—Beech. Gaelic and Irish : *craobh fhaibhile.* Welsh : *ffawydd. Fai,* from φαγω, to eat. φηγός, the beech-tree. This name was first applied to the oak, and as we have no *Quercus esculus,* the name *Fagus* is applied to the beech and not to the oak. *Oruin* (O'Reilly), see *Thuja articulata. Beith na measa,* the fruiting birch. *Meas,* a fruit, as of oak or beech—like "mess," "munch." French : *manger,* to eat.

F. sylvatica var. **atrorubens**—Black beech. Gaelic : *faibhile dubh* (Fergusson), black beech, from the sombre appearance of its branches. The "mast" of the beech was used as food, and was called *bachar,* from Latin, *bacchar ;* Greek, βάκχαρις, a plant having a fragrant root. A name also given to *Valeriana celtica* (Sprengel), Celtic nard.

Carpinus—Celtic : *car,* wood ; and *pin,* a head. It having been used to make the yokes of oxen.

C. betulus—Hornbeam. Gaelic : *leamhan bog*, the soft elm. (See *Ulmus campestris*).
Corylus avellana—Hazel. Gaelic and Irish : *càlltuinn, càll-dainn, càllduinn, cailtin, colluinn*. Welsh : *callen*. Cornish : *col-widen*. Perhaps from Armoric : *call*. Gaelic : *coill*. Irish : *coill*, a wood, a grove. New Year's time is called in Gaelic, *coill ; " oidhche coille,"* the first night of January, then the hazel is in bloom. The first night in the new year, when the wind blows from the west, they call *dàir na coille*, the night of the fecunda-tion of trees ("Statistics," par. Kirkmichael). In Celtic supersti-tion the hazel was considered unlucky, and associated with loss or damage. The words *càll, còl, collen*, have also this significa-tion ; but if two nuts were found together (*cnò chòmhlaich*), good luck was certain. The Bards, however, did not coincide with these ideas. By it they were inspired with poetic fancies. " They believed that there were fountains in which the principal rivers had their sources : over each fountain grew nine hazel trees, *caill crinmon* (*crina*, wise), which produced beautiful red nuts, which fell into the fountain, and floated on its surface, that the salmon of the river came up and swallowed the nuts. It was believed that the eating of the nuts caused the red spots on the salmon's belly, and whoever took and ate one of these salmon was inspired with the sublimest poetical ideas. Hence the ex-pressions, ' the nuts of science,' ' the salmon of knowledge.' " O'Curry's ' Manners and Customs of the Ancient Irish.'
The badge of Clan Colquhoun.
Alnus—Name derived from Celtic. *Al*, a growth ; and *lan*, full. According to Pictet, it is from *alka*, Sanskrit for a *tree*.
A. glutinosa—Common alder. Gaelic and Irish : *feàrn—feàrn*, same origin as *varâna* (Sanskrit), a tree. Welsh : *gwernen —gwern*, a swamp. It grows best in swampy places, and beside streams and rivers. Many places have derived their names from this tree, *Gleann Fearnaite*. *Fearnan*, near Loch Tay ; *Fearn*, Ross-shire, &c. *Ruaim* (O'Reilly) (*ruadh*, red), it dyes red. When peeled it is white, but it turns red in a short time. The bark boiled with copperas makes a beautiful black colour. The wood has the peculiarity of splitting best from the root, hence the saying

" Gach fiodh o'n bhàrr, 's am *fairna* o'n bhun."
Every wood splits best from the top, but the alder from the root.

Betula alba—Birch. Gaelic and Irish : *beatha*. Welsh :

bedu, seemingly from *beath*. Greek: βιωτη. Latin: *vita*, life. Also the name of the letter *B* in Celtic languages, corresponding to Hebrew *Beth* (meaning a house). Greek: *Beta*. Generally written *beith*.

> " Sa *bheith* chubhraidh."—OSSIAN.
> In the fragrant birch.

The Highlanders formerly made many economical uses of this tree. Its bark (*meilleag*), they burned for light, and the smooth inner bark was used, before the invention of paper, for writing upon, and the wood for various purposes.

The badge of the Clan Buchanan.

B. verrucosa—Knotty birch. Gaelic: *beatha carraigeach*, the rugged birch; *beatha dubh-chasach*, the dark-stemmed birch.

B. pendula—Gaelic: *beatha dubhach*, the sorrowful birch (*dubhach*, dark, gloomy, sorrowful, mourning, frowning). In Rannoch and Breadalbane: *Beatha cluasach*, the many (drooping) *ear* birch. (Stuart.)

B. nana.—Dwarf birch. Gaelic: *beatha beag* (Fergusson), the small birch.

Castanea vesca—Common chestnut. Gaelic and Irish: *chraobh geanm-chno*.

> "No na craobha *geanm-chno* cosmhuil r'a gheugaibh."—EZEKIEL xxxi. 8.
> Nor the chestnut-tree like his branches.

Geanm or *geanm*, natural love, pure love, such as exists between relatives,—the tree of chaste love, and *cno*, a nut. The Celts evidently credited this tree with the same virtues as the chaste tree, *Vitex agnus castus* (Greek, ἀγνὸς; and Latin, *castus*, chaste). Hence the Athenian matrons, in the sacred rites of Ceres, used to strew their couches with its leaves. *Castanea* is said to be derived from Castana, a town in Pontus, and that the tree is so called because of its abundance there. But the town Castana (Greek, Κάστανον) was probably so called on account of the virtues of its female population. If so, the English name chestnut would mean chaste-nut, as it is in the Gaelic. Welsh: *castan* (from Latin, *caste*), chastely, modestly. The chestnut-tree of Scripture is now supposed to be *Platanus orientalis*, the Chenar plane-tree.

[**Æsculus hippocastanum** — The horse-chestnut. Gaelic: *geanm chno feadhaich* (Fergusson). Belongs to the order *Aceraceæ*. Was introduced to Scotland in 1709.]

Populus alba — Poplar. Gaelic : *pobhuill.* Irish : *poibleag.* German : *pappel.* Welsh and Armoric : *pobl.* Latin : *populus.* This name has an Asiatic origin, and became a common name to all Europe through the Aryan family from the East.[1] Pictet explains it thus : "Ce nom est sans doute une reduplication de la racine Sanskrit *pul,* magnum, altum." *Pul pul,* great, great, or big, big, as in the Hebrew construction, very big. We still say in Gaelic *mòr mòr,* big, big, for very big. *Pul pul* is the Persian for poplar, and *pullah* for salix. This tree is quite common in Persia and Asia Minor, hence it was as well known there as in Europe. The name has become associated with *populus,* the people, by the fact that the streets of ancient Rome were decorated with rows of this tree, whence it was called *Arbor populi.* Again, it is asserted that the name is derived from the constant movement of the leaves, which are in perpetual motion, like the *populace*—" fickle, like the multitude, that are accursed."

P. tremula—Aspen. Gaelic and Irish : *critheann,* trembling.

> " Mar *chritheach* san t' sine."—ULL.
>
> Like an aspen in the blast.

With the slightest breeze the leaves tremble, the poetic belief being that the wood of the Cross was made from this tree, and that ever since the leaves cannot cease from trembling. *Eadhadh.* Welsh : *aethnen (aethiad,* smarting). The mulberry tree of Scripture is supposed to be the aspen (Balfour), and in Gaelic is rendered *craobh nan smèur.* (See *Morus* and *Rubus fruticosus.*)

> "Agus an uair a chluineas tu fuim siubhail an mullach *chraobh nan smèur,* an sin gluaisidh tu thu féin."—2 SAMUEL v. 24.
>
> And when thou hearest a sound of marching on the tops of the mulberry trees, that then thou shalt bestir thyself.

The badge of Clan Fergusson.

Salix—According to Pictet, from Sanskrit, *sâla,* a tree.

> " Il a passe au *suale* dans plusieurs langues
> . . . Ces noms derivent de sâla."

Gaelic and Irish : *seileach, saileog, sal, suil.* Cognate with Latin : *salix.* Fin.: *salawa.* Anglo-Saxon : *salig, salh,* from which

---

[1] See Canon Bourke's work on 'The Aryan Origin of the Gaelic Race and Language.' London : Longman.

sallow (white willow) is derived. Welsh: *helyg*, willow. (See *S. viminalis*.)

S. viminalis — Osier willow; cooper's willow. Gaelic and Irish: *fineamhain* (from *fin*, vine; and *muin*, a neck), a long twig—a name also applied to the vine.[1] *Vimen* in Latin means also a pliant twig, a switch osier. One of the seven hills of Rome (Viminalis Collis) was so named from a willow copse that stood there; and Jupiter, who was worshipped among these willows, was called "Viminius;" and his priests, and those of Mars, were called *Salii* for the same reason. The worship was frequently of a sensual character, and thus the willow has become associated with lust, filthiness. Priapus was sarcastically called "Salacissimus Jupiter," hence *salax*, lustful, salacious; and in Gaelic, *salach* (from *sal*); German, *sal*, polluted, defiled. The osier is also called *bunsag*, *bun*, a stump, a stock. *Maothan*, from *maoth*, smooth, tender. *Gall sheileach*, the foreign willow.

S. caprea, and **S. aquatica**—Common sallow. Gaelic and Irish: *sùileag*, probably the same as Irish, *saileog* (Anglo-Saxon, *salig*, sallow). *Sùil*—the old Irish name—(in Turkish *su* means water) in Irish and Gaelic, the eye, look, aspect, and sometimes *tackle* (Armstrong). The various species of willow were extensively used for tackle of every sort. Ropes, bridles, &c., were made from twisted willows. "In the Hebrides, where there is so great a scarcity of the tree kind, there is not a twig, even of the meanest willow, but what is turned by the inhabitants to some useful purpose."—WALKER's 'Hebrides.' And in Ireland to this day "gads," or willow ropes, are made. *Geal-sheileach* (Armstrong), the white willow or sallow tree. Irish: *crann sailigh fhrancaigh*, the French willow.

S. babylonica—The Babylonian willow. Gaelic: *seileach an t' srutha* (*sruth*, a brook, stream, or rivulet), the willow of the brook.

"Agus gabhaidh sibh dhuibh féin air a' cheud là meas chraobh àluinn, agus *seileach an t' srutha*."—LEV. xxiii. 40.

And take unto yourselves on the first day fruit of lovely trees, and willows of the brook.

MYRICACEÆ.

Myrica gale—Bog myrtle, sweet myrtle, sweet gale. Gaelic: *rideag*. Irish: *rideog*, *rileog* (changing sound of *d* to *l* being

[1] "*Finemhain* fa m' chomhair" (in Genesis)—a vine opposite to me.

easier). *Ròd* or *roid* is the common name in the Highlands, perhaps from the Hebrew, רחם, *rothem*, a fragrant shrub. It is used for numerous purposes by the Highlanders, *e.g.*, as a substitute for hops; for tanning; and from its supposed efficacy in destroying insects, beds were strewed with it, and even made of the twigs of gale, which is there called *nodha*. "And to this day it is employed by the Irish for the same purpose by those who know its efficacy. The *rideog* is boiled and the tea or juice drank by children to kill 'the worms.' I think children educated in our national schools should be taught to know these plants and their value."—CANON BOURKE.

Badge of the Clan Campbell.

CONIFERÆ.

Pinus—French : *le pin*. German : *pyn-baum*. Italian : *il pino*. Spanish : *el pino*. Irish : *pinn chrann*. Gaelic : *pin - chrann*. Anglo-Saxon : *pinu*. All these forms of the same name are derived, according to Pictet, from the Sanskrit verb *pina*, the past participle of *pita*, to be fat, juicy. From *pina*, comes Latin, *pinus*, and the Gaelic, *pin*.

P. **sylvestris** — Scotch pine, Scots fir. Gaelic : *giùthas*, *giùbhas*.

> " Mar *giùbhas* a lùb an doinionn."—OSSIAN.
>
> Like a pine bent by the storm.

Giùthas, probably from the same root as *picea*, pitch pine. Sanskrit : *pish*, soft, juicy. Gaelic : *giùbhas*, a juicy tree,—from the abundance of pitch or resin its wood contains; *Con* or *cona* (O'Reilly), from Greek : χωνος, *konos*, a cone, a pine. Hence *conadh*, fire-wood. *Fir* in English, from Greek, πῦρ, fire, because good for fire.

Badge of the Macgregors—Clan Alpin.

P. **picea**—Silver pine. Gaelic : *giùbhas geal* (Fergusson), white pine. First planted at Inveraray Castle in 1682.

Abies communis—Spruce-fir. Gaelic : *guithas Lochlannach*, Scandinavian pine.

> " Nuair theirgeadh *giùbhas Lochlainneach*."—M'CODRUM.
>
> When the spruce fir is done.

Lòchlannach, from *loch*, lake, and *lann*, a Germano-Celtic word meaning land—*i.e.*, the lake-lander, a Scandinavian.

" Giubhas glàn na Lòchlainn,
Fuaight' le copar ruadh."

Polished fir of Norway,
Bound with reddish copper.

P. larix—Larch. Gaelic and Irish : *laireag.* Scotch : *larick.*
Latin : *larix*, from the Celtic, *làr*, fat, from the abundance of
resin the wood contains. Welsh : *larswydden*, fat wood.

P. strobus—(*Strobus*, a name employed by Pliny for an eastern tree used in perfumery) Weymouth pine. Gaelic : *giùthas
Sasunnach* (Fergusson), the English pine. It is not English,
however ; it is a North American tree, but was introduced from
England to Dunkeld in 1725.

Cupressus—Cypress. Irish and Gaelic : *cuphair*, an alteration of Cyprus, where the tree is abundant.

C. sempervirens—Common cypress. Gaelic : *craobh bhròin*,
the tree of sorrow. *Bròn*, grief, sorrow, weeping. *Craobh uaine
giùthais*, the green fir-tree.

> " Is cosmhuil mi ri *crann uaine giuthais*."—HOSEA xiv. 8.
> I am like a green fir-tree.

The fir-tree of Scripture (Hebrew *berosh* and *beroth* are translated
fir-trees) most commentators agree is the cypress.

Thuja articulata—Thyine wood. Gaelic : *fiodh-thine.*

> " Agus gach uile ghnè *fhiodha thine*."—REV. xviii. 12.
> And all kinds of thyine wood.

Alteration of *thya*, from θυω, to sacrifice. Another kind of
pine, Hebrew, *oren* (Irish and Gaelic, *oruin*), is translated ash
in Isaiah xliv. 14, and beech by O'Reilly.

Cedar — (So called from its firmness.) Hebrew : אֶרֶז, *crez.*
Cedrus Libani, cedar of Lebanon. Gaelic and Irish : *crann
sheudar*, cedar-tree.

> " Agus air uile *sheudaraibh Lebanoin*."—ISAIAH ii. 13.
> And upon all the cedars of Lebanon.

The *cedar wood* mentioned in Lev. xiv. 4, was probably *Juniperus
oxycedrus*, which was a very fragrant wood, and furnished an oil
that protects from decay—cedar oil (κέδριον). " Carmina linenda
cedro "—*i.e.*, worthy of immortality.

> " Agus *fiodh sheudar*, agus scàrlaid, agus hiosop."
> And cedar wood, scarlet, and hyssop.

K

Juniperus—Said to be "from the Celtic *jeneprus*, which sig-
nifies rough or rude" (Loudon), a word *not* occurring in any
Celtic vocabularies that I have consulted. It seems to be the
Latinised form of the Celtic root *iu, iubh, iur, yw* (see *Taxus*).
From the same root comes *yew* in English. Irish: *iubhar-
beinne* (O'Reilly), the hill yew; *iubhar-talamh*, the ground yew;
ubhar-chraige, the rock yew; all given as names for the juniper.
Juniperus is mentioned by both Virgil and Pliny. Both the
Greeks and Romans reluctantly admitted that they were indebted to the Celts for many of their useful sciences, and even
their philosophy (see Diogenes Laertius), as they certainly were
for their plant and geographical names.

J. communis—Juniper. Gaelic and Irish: *aiteil, aitinn,
aitiol.*

"Ach chaidh e fèin astar làtha do'n fhasach agus thàinaig e agus shuidh
e fuidh *craobh aiteil.*"—1 KINGS xix. 4.

And he went a day's journey into the desert, and he sat under a juniper
tree.

The juniper of Scripture, *Genista monosperma*, was a kind of
broom. *Aiteil*, from *ait*. Welsh: *aeth*, a point, furze. Irish:
aiteann, furze, from its pointed leaves. *Bior leacain* (in Arran),
the pointed hill-side plant. *Staoin* (in the North Highlands),
caoran staoin, juniper berries (*staoin*, a little drinking-cup).

The badge of Clans Murray, Ross, M'Leod, and the Athole
Highlanders.

J. sabina—Savin. Gaelic: *samhan* (Armstrong), alteration of
"sabina" the "sabina herba" of Pliny. Common in Southern
Europe, and frequently cultivated in gardens, and used medicinally as a stimulant, and in ointments, lotions, &c.

Taxus—According to Benfey is derived from the Sanskrit,
taksh, to spread out, to cut a figure, to fashion. Persian *tak*.
Greek: *τοξος*, an arrow. Irish and Gaelic: *tuagh*, a bow made
of the *taxos* or yew, now applied to the hatchet used in place of
the old bow.

T. baccata — Common yew. Gaelic and Irish: *iuthar,
iubhar, iughar*, from *iùi*. Greek: *ιός*, an arrow, or anything
pointed. Arrows were poisoned with its juice; hence in old
Gaelic it was called *iogh*, a severe pain, and *ioghar* (Greek,
ιχωρ, ichor) pus, matter. The yew was the wood from which
ancient bows and arrows were made, and that it might be ready
at hand, it was planted in every burial-ground.

" 'N so fein, a Chuchullin, tha' n ùir,
'S caoin *iuthar* 'tha 'fàs o'n uaigh." [1]—OSSIAN.

In this same spot Chuchullin, is their dust,
And fresh the yew tree grows upon their grave.

Hence another form of the name *eo*, a grave. *Sìnsior, sinnsior* (O'Reilly), long standing, antiquity, ancestry. The yew is remarkable for its long life. The famous yew of Fortingall in Perthshire, which once had a circumference of 56½ feet, is supposed to be 3500 years old. *Sineadhfeadha* (O'Reilly), protracting, extending.

The badge of Clan Fraser.

ENDOGENS.

ORCHIDACEÆ.

Orchis—Greek: ὄρχις, a plant with roots in the shape of testicles. "Mirabilis est *orchis* herba, sive serapias, gemina radice testiculis simili"—PLINY.

O. maculata—The spotted orchis. Gaelic and Irish: *ùrach bhallach*, from *ùr*, fresh; *ùrach*, a bottle; *uradh*, apparel, and *ballach*, spotted.

O. mascula—Early orchis. Gaelic: *moth-ùrach*, from *mòth*, the male of any animal.

" Lointeann far an cinn
I'na *moth'raichean*."—M'INTYRE in ' Ben Doran.'

Meadows where the early orchis grow.

Irish: *magairlin meireach*, (*magairle*, the testicles; *meireach* (Greek, *meiro*), joyful, glad). *Clachan gadhair* (*gadhar* a hound, *clach*, a stone). The name, *cuigeal an losgain*, the frog's spindle, is applied to many of the orchis; and frequently the various names are given to both *maculata* and *mascula*.

O. conopsea — Fragrant orchis. Gaelic: *lus tàghta*, the chosen or select weed.

Ophrys—Greek: οφρύς (Gaelic, *abhra*), the eyelash, to which the delicate fringe of the inner sepals may be well compared. "A plant with two leaves"—FREUND.

[1] Laing is not correct when, in attacking the genuineness of the poems of Ossian, he asserts that the yew, so often mentioned in these poems, is not indigenous. There are various places, such as Gleniur, Duniur, &c., that have been so named from time immemorial, which prove that the yew was abundant in these places at least many centuries ago.

O. or Listera ovata—Tway blade. Gaelic: *dà-dhuilleach*, two-leaved; *dà-bhileach*, same meaning.

Epipactis latifolia—White helleborine. Gaelic: *'elebor-geal.*[1] A plant used formerly for making snuff. "The root of hellebor cut in small pieces, the pouder drawne vp into the nose causeth sneezing, and purgeth the brain from grosse and slimie humors" —GERARD, 1597. This is probably the plant referred to in "Morag," when M'Donald describes the buzzing in his head, for even his nose he had to stop with *hellebore*, since he parted from her endearments.

> " Mo cheann tha làn do sheilleanaibh
> O dheilich mi ri d'bhriodal
> Mo shròn tha stoipt' á *dh-elebor.*
> Na deil, le teine dimbis."

IRIDACEÆ.

Iris—Signifying, according to Plutarch, the "eye." Canon · Bourke maintains " it is derived from εἰρω, to settle. And as a name it was by the pagan priests applied to the imaginary messenger, sent by gods and goddesses to others of their class, to announce tidings of goodwill. At times they imagined her sent to mortals, as in Homer, *to settle* matters, or to say they were destined to be settled. Such was the duty of IRIS. Now amongst Jews and Christians, the rainbow was the harbinger of *peace* to man, hence it was called ' Iris ; ' and the circle of blue, grey, or variegated tints around the pupil of the eye is not unlike the rainbow—therefore this circlet was so called by optic scientists, simply because they had no other word ; and botanists have, by comparison, applied it to the *fleur-de-lis*, because it is varied in hue, like the iris of the eye, or the rainbow. *Iris* does not and did not convey the idea of eye."

I. pseudacorus — The yellow flag. Gaelic : *bog-uisge* — *bog*, soft, but here a corruption of *bogha-uisge*, the rainbow. Gaelic and Irish : *seilisdear*, often *seileasdear*, and *siolastar*. The termination, *tar, dear*, or *astar*, in these names, means one of a kind, having a settled form or position. One finds this ending common in names of plants—as, *oleaster, cotoneaster*, &c., like " τηρ " in Greek, "fear " in Gaelic. *Seil* (the first syllable), from *sol*, the sun ; *solus*, light ; *sol* and *leus, i.e., lux*, light. Greek : ῝Ηλιος (η or *e* long), hence *sēil, e* and *i* to give a lengthened sound,

---

[1] See *Helleborus viridis.*

as in Greek. *Seileastar*, therefore, means the plant of light –
Fleur de luce. Other forms of the word occur. *Siol* instead
of *seil*, as *siolstrach ; siol* or *sil*, to distil, to drop—an alteration
probably suggested by the medicinal use made of the roots of
the plant, which were dried, and made into powder or snuff,
to produce salivation by its action on the mucous membrane.
" *Feileastrom, feleastrom, feleastar.* Here *f* is the affected or digammated form. When *eleastar* (another form of the word) lost
the '*s*,' then, for sound's sake, it took the digammated form
(*f*)*eleastar. Strom* (the last syllable) is a diminutive termination. *Seilistear*, diminutive form *seilistrin*, and corrupted into
seilistrom "—BOURKE.

Crocus—Greek : κρόκος. Much employed amongst the ancients for seasonings, essences, and for dyeing purposes.

C. sativus }
Colchicum autumnale } —Saffron crocus, meadow saffron.
Gaelic and Irish : *crò, cròdh, cròch—cròdh chorcar.*[1]

> " 'Se labhair Fionn nan chrò-shnuaidh."—CONN MAC DEARG.
> Thus spake Fingal the saffron-hued.

> " Spiocnard agus croch."—DANA SHOLIIIM, iv. 14.
> Spikenard and saffron.

Saffron was much cultivated anciently for various purposes, but
above all for dyeing. " The first habit worn by persons of distinction in the Hebrides was the *lein croich*, or saffron shirt, so
called from its being dyed with saffron."—WALKER. The Romans
had their *crocōta*, and the Greeks ὁ κροκωτός, a saffron-coloured
court dress. Welsh : *saffrwm*, saffron, from the Arabic name,
z'afarân, which indicates that the name of the plant is of Asiatic
origin.

AMARYLLIDACEÆ.

Narcissus pseudo-narcissus }
„   **jonquilla** } —Daffodil. Gaelic : *lus a chròmchinn*, the plant having a bent or drooping head.

Galanthus nivalis — Snowdrop. Gaelic and Irish : *gealag
lair, —gealag*, white as milk ; *làr*, the ground. *Galanthus.*
Greek : γάλα, milk, and ἄνθος, a flower.

Aloe—Hebrew, אהלות, *ahaloth.* Gaelic and Irish : *aloe.*

> " Leis na h-uile chraobhaibh tuise, mirr agus *aloe.*"
> With all trees of frankincense ; myrrh, and aloes.—SONG OF SOLOMON,
> iv. 14.

---

[1] For *corcur*, see *Lecanora tartarea.*

The aloe of Scripture[1] must not be confounded with the bitter herb well known in medicine.

LILIACEÆ.

Lilium—Greek : λείριον. From the Celtic : *li*, colour, hue. Welsh : *lliu.* Gaelic : *li.*

"A mhaise-mhna is ailidh *li!*"—FINGALIAN POEMS.
Thou fair-faced beauty.

"Lily seems to signify a flower in general." — WEDGEWOOD. Gaelic and Irish : *lilidh* or *lli.*

Convallaria majalis—Lily of the valley. Gaelic : *lili nan lòn. Lili nan gleann.*

"Air ghilead, mar *lili nan lòintean.*"—M'DONALD.
White as the lily of the valley.
"Is ròs Sharon mise *lili nan gleann.*"—STUART.
I am the rose of Sharon, the lily of the glen.

"The lily of Scripture was probably *Lilium chalcedonicum.*"— BALFOUR.

Allium—The derivation of this word is said to be from *all* (Celtic), hot, burning. There is no such word. The only word that resembles it in sound, and with that signification, is *sgallta*, burned, scalded; hence, perhaps, "scallion," the English for a young onion. Latin : *calor.*

A. cepa (*cep*, Gaelic : *ceap*, a head) — The onion. Gaelic : *uinnean.* Irish : *oinninn.* Welsh : *wynwyn.* French : *oignon.* German : *önjön.* Latin : *unio.* Gaelic : *siobaid, siobann.* Welsh : *sibol.* Scotch : *sybo.* German : *zwiebel*, scallions or young onions. *Cutharlan*, a bulbous plant. In Lorne, and elsewhere along the W. Highlands, frequently called *Srònamh* (probably from *Sròn* and *amh, raw* in the *nose*, or *pungent* in the *nose*).

A. porrum[2]—Garden leek. Gaelic and Irish : *leigis, leiceas, leicis.* German : *lauch*, leek.

"Agus na *leicis* agus na h'*uinneinean.*"—NUMBERS, xi. 5.
And the leeks and the onions.

Irish : *bugha* (Shaw), leeks, fear. O'Clery, in his 'Vocabulary,' published A.D. 1643, describes it thus : "*Bugh, i.e.*, luibh gorm nó glàs ris a samhailtean sùile bhios gorm no glàs." That is, a blue or grey plant, to which the eye is compared if it be blue or

---

[1] *Aquilaria agallochum.*
[2] "Porrum" from the Celtic, *pori*, to eat, to graze, to browse.

grey. The resemblance between a leek and the eye is not very apparent, as the following quotation shows :—

"Dhearca mar dhlaoi don *bhugha*,
Is a dha bhraoi cearta caol-dhubha."—O'BRIEN.

His eyes like a bunch of leeks,
And his two eyebrows straight, dark, narrow.

Although Shaw gives the name to leek, probably the plant referred to is the harebell (see *Scilla non - scripta*). Irish : *coindid, coinne, cainnen.* Welsh : *cenin* (*cen*, a skin, peel, scales, given to onions, garlic, leeks).

"Do roidh, no do *coindid*, no do ablaibh."
Thy gale, nor thy onions, nor thy apples.

Coindid, though applied to leeks, onions, &c., means seasoning, condiments, Latin : *condo.*

A. ursinum — Wild garlic. From the Celtic. Gaelic and Irish : *garleag.* Welsh : *garlleg*, from *gar, gairce*, bitter, most bitter. *Gairgean. Creamh* (Welsh, *craf*), *cream*, to gnaw, chew. *Lurachan*, the flower of garlic.

"Le d' *lurachain chreamhach* fhàson
· 'Sam buicein bhàn orr' shuas."—M'DONALD.

The feast of garlic, " Fèisd chreamh," was an important occasion for gatherings and social enjoyment to the ancient Celts.

"Ann's bidh creamh agus sealgan, agus luibhe iomdha uile fhorreas, re a n-itheadh ùrghlas feadh na bleadhna ma roibhe ar teitheadh ó chaidreath na n-daoine, do 'n gleann dà loch."—IRISH.

Where garlic and sorrel, and many other kinds, of which I ate fresh throughout the year before I fled from the company of men to the glen of the Two Lochs.[1]

A. scorodoprasum — Rocambole. Gaelic and Irish : *creamh nan crag* (M'Kenzie), the rock garlic.

A. ascalonicum—Shallot. Gaelic : *sgalaid* (Armstrong). (See *Allium*).

A. schœnoprasum—Chives. Gaelic : *feuran.* Irish : *fearan*, the grass - like plant. *Saidse. Creamh ghàradh*, the garden garlic. Welsh : *cenin Pedr*, Peter's leek.

A. vineale — Crow garlic. Gaelic : *garleag Mhuire* (Armstrong), Mary's garlic.

[1] A most gloomy and romantic spot in the County of Wicklow.

"Glen da lough ! thy gloomy wave,
Soon was gentle Kathleen's grave."--MOORE.

Narthecium ossifragum—Bog asphodel. Gaelic and Irish : *blioch, bliochan,* from *blioch,* milk. Welsh : *gwaew'r trenin,* king's lance.

> " Nuair thigheadh am buaichaill a mach,
> 'Sa gabhadh e mu chùl a chrùidh
> Mu'n cuairt do Bhad-nan-clach-glas,
> A bhuail 'air m bu tric am *bliochd.*"—M'LEOD.

> When the cowherd comes forth,
> And follows his cows
> Around Bhad-nan-clach-glàs,
> Often he is struck with the asphodel.

Scilla non-scripta—Bluebell ; wild hyacinth. Gaelic : *fuath mhuic,* the pig's fear or aversion, the bulbs being very obnoxious to swine. *Brog na cubhaig,* cuckoo's shoe. Irish : *buth a muc.* Probably *buth* is the same as *bugha* (see *Allium porrum*), fear, the pig's fear. M'Lauchainn called it *lili gucagach.*

> " *Lili gucagach* nan cluigean."
> The bell-flowered lily.

S. verna — Squill (and the Latin, *scilla,* from the Arabic, *ăsgyl*). Gaelic : *lear uineann,* the sea-onion. *Lear,* the sea, the surface of the sea.

> "Clos na min-*lear* uaine."—OSSIAN.
> The repose of the smooth green sea.

Welsh : *winwyn y mor,* sea-onion.

Tulipa sylvestris—Tulip. Gaelic : *tuiliop.* The same name in almost all European and even Asiatic countries. Persian : *thoùlybân* (De Souza).

Asparagus officinalis—Common asparagus. Gaelic : *creamh mac-fiadh.* Irish : *creamh-mùicfiadh,* wild boar's leek or garlic. The same name is given to hart's tongue fern. *Aspàrag,* from the generic name σπαρασσω, to tear, on account of the strong prickles with which some of the species are armed.

Ruscus—Latinised form of Celtic root *rus,* wood, husk ; *rusgach,* holly. Welsh : *rhysgiad,* an over-growing. Also *bruscus,* from Celtic, *brus, bruis,* small branches, brushwood.

R. aculeatus — Butcher's broom. Gaelic : *calg-bhrudhainn* (Armstrong). Irish : *calg-bhrudhan* (Shaw)--*calg,* a prickle, from its prickly leaves ; and *bruth, bruid,* a thorn, anything pointed ; *brudhan,* generally spelled *brughan,* a faggot. Or it may only be a corruption from *brum,* broom. *Calg bhealaidh,* the prickly broom. It was formerly used by butchers to clean their blocks, hence the English name " butchers' broom."

NAIADACEÆ.

Potamogeton.— Greek : ποταμός, a river, and γείτον, near. **P. natans** — Broad-leaved pondweed. Gaelic: *duiliasg na h'aibhne*, the river leaf. Most of the species grow immersed in ponds and rivers, but flower above its surface. *Liobhag*, from *liobh*, smooth, polish, from the smooth pellucid texture of the leaves, their surface being destitute of down or hair of any kind. Irish : *liachroda,—liach*, a spoon, *rod*, a water-weed, sea weed ; *liach-Brighide*, Bridget's spoon. Probably these names were also given to the other species of pondweeds (such as *P. polygonifolius*) as well as to *P. natans.*

Zostera marina—The sweet sea-grass. Gaelic and Irish : *bilearach* (in Argyle, *bileanach*), from *bileag*, a blade of grass. The sea-grass was much used for thatching purposes, and it was supposed to last longer than straw.

ALISMACEÆ.

Alisma.—Greek : ἄλισμα, an aquatic plant ; said to be from a Celtic root, *alis*, water. If ever this was a Celtic vocable it has ceased to have this signification : in Welsh *alis* means the lowest point, hell.

A. Plantago—Water-plantain. Gaelic and Irish : *cor-chopaig* (*cor* or *cora*, a weir, a dam, and *copag*, a dock, or any large leaf of a plant). It grows in watery places. Welsh : *llyren*, a duct, a brink or shore.

Triglochin palustre—Arrow-grass. Gaelic : *bàrr a' mhilltich,*—

"Bun na clpe is *bàrr a' mhilltich.*"—M'INTYRE.

bàrr, top, and *milltich* (Irish), "good grass," and *milneach*, a thorn or bodkin—hence the English name arrow-grass. Generic name from τρεῖς, three, and γλωχίς, a point, in allusion to the three angles of the capsule. Sheep and cattle are fond of this hardy species, which afford an early bite on the sides of the Highland mountains. *Milltich* is commonly used in the sense of "grassy;" *maghanan millteach*, verdant or grassy meadows.

LEMNACEÆ.

Lemna minor—Duckweed. Gaelic :[1] *mac gun athair*, son without a father. Irish : *lus gan athair gan mhathair*, fatherless motherless wort. A curious name, perhaps suggested by the

[1] *Mac-gun-athair* may have originally been *meacan air,—meacan*, a plant, *air*, gen. of *àr*, slow (hence the name of the river "Arar" in France, meaning the slow-flowing river,—"*Arar* dubitans qui suos cursos agat"— SENECA), the plant that grows in slow or sluggish water.

L

root being suspended from its small egg-shaped leaf, and not affixed to the ground. *Gran-lachan,—gran,* seed, grain, and *lach,* a duck. The roundish leaves, and the fact that ducks are voraciously fond of feeding on them, have suggested this and the following names : *Ròs lachain,* the ducks' rose or flower. Irish : *abhran donog* (O'Reilly),—*abhran* is the plural of *abhra,* an eyelid, and *donog,* a kind of fish, a young ling. The fish's eyelids ; more likely a corruption of *aran tunnaig,* duck's bread or meat.

It was used by our Celtic ancestors as a cure for headaches and inflammations.

ARACEÆ.

Arum, formerly *aron,* probably from the ancient Celtic root *ar,* land, earth ; hence Latin, *aro,* to plough, and Gaelic, *aran,* bread, sustenance. The roots of many of the species are used both for food and medicine.

A. maculatum—Wake-robin, lords and ladies. Gaelic : *cluas chaoin,* the soft ear (*caoin,* soft, smooth, gentle, &c., and *cluas,* ear). The ear-shaped spathe would probably suggest the name. *Cuthaidh,* from *cuth,* a head, a bulb—hence *cutharlan,* any bulbous-rooted plant. *Cuthaidh* means also wild, savage. *Gachar* and *gaoicin cuthigh* are given in O'Reilly's Dictionary as names for the Arum, from *cai,* a cuckoo. Old English : cuckoo's pint.

ORONTIACEÆ.

Acorus calamus—Sweet-flag. Gaelic : *cuilc-mhilis,* sweet-rush ;

" *Cuilc mhilis* agus canal."

Calamus and cinnamon.

cuilc, a reed, a cane. Greek : κάλαμος, applied to reeds, bulrush canes, *e.g., cuile na Léig,* the reeds of Lego. *Cobhan cuilc,* an ark of bulrushes. *Cuilc-chrann,* cane ; *milis* (Greek : μέλισσα, a bee), sweet. Before the days of carpets, this plant is said to have supplied the "rushes" with which it was customary to strew the floors of houses, churches, and monasteries.

TYPHACEÆ.

Typha, from Greek, τυφος, a marsh in which all the species naturally grow.

T. latifolia—Great reed-mace or cat's-tail. Gaelic and Irish : *bodan dubh,* from *bod,* a tail, and *dubh,* large, or dark. *Cuigeal nam bàn-sìth,* the fairy-woman's spindle. It is often, but incorrectly, called *bog bhuine* or bulrush (see *Scirpus lacustris*). The downy seeds were used for stuffing pillows, and the leaves for making mats, chair - bottoms, thatch, and sometimes straw hats or bonnets.

T. angustifolia—Lesser reed-mace or cat's-tail. Irish : *bodan* (O'Reilly), dim. of *bòd*, a tail.

Sparganium.—Name in Greek denoting a little band, from the ribbon-like leaves.

S. ramosum—Branched bur-reed. Gaelic: *righ seisg*, the king's sedge, from its being a large plant with sword-shaped leaves. *Seisg mheirg* (Stewart),—*meirg*, rust, a standard or banner.

S. simplex— Upright bur - reed. Gaelic: *seisg madraidh*. Armstrong gives this name to *S. erectum*, by which he doubt-less means this plant. *Seisg*, sedge, and *madradh*, a dog, a mastiff. Name probably suggested by the plant being in per-fection in the dog-days, the month of July, *mìos Mhadrail.*

JUNCACEÆ.

Juncus, from the Latin *jungo*, to join. The first ropes were made from rushes, and also floor covering. Ancient Gaelic : *aoin*, from *aon*, one. Latin : *unus.* Greek : εν. Ger.: *ein.*

> " A dàth amar dhàth an aeil,
> Coilcigh eturra agus *aein.*
> Sída eturra is brat gorm,
> Derg òr eturra is glan chorn."

(From the description of the Lady Crehé's house by Caeiltè MacRonain, from the Books of Ballymote, a rare ancient poem.)

> The colour [of her *dùn*] is like the colour of lime :
> Within it are couches and *green rushes ;*
> Within it are silks and blue mantles ;
> Within it are red gold and crystal cups.

J. conglomeratus—Common rush. Gaelic and Irish : *luachar*, a general name for all the rushes, meaning splendour, brightness; hence *luachar*, a lamp. Latin: *lucerna.* Sanscrit: *lauchanan*, from the root, *lauch*, light. The pith of this and the next species was ·commonly used to make rush-lights. The rushes were stripped of their outer green skin, all except one narrow stripe, and then they were drawn through melted grease and laid across a stool to set. "The title *Luachra* was given to the chief Druid and magician, considered by the pagan Irish as a deity, who opposed St Patrick at Tara in the presence of the king and the nobility, who composed the convention"—'Life of St Patrick.' *Bròg braidhe* (O'Reilly),—*bròg*, a shoe ; but here it should be *bròdh*, straw ; *braidhe*, a mountain, the mountain straw or stem.

J. effusus—Soft rush. Gaelic: *luachar bòg*, soft rush. Irish: *feath*, a bog. It grows best in boggy places. *Fead*, which seems to be the same name, is given also to the bulrush. *Fead*, a whistle, a bustle.

> " 'S lìonmhor *feadan* caol,
> Air an éirich gaoth."—M'INTYRE.

Doubtless suggested by the whistling of the wind among the rushes and reeds. The common rush and the soft rush were much used in ancient times as bed-stuffs; they served for strewing floors, making rough couches, &c.

J. articulatus—Jointed rush. Gaelic: *lochan nan damh.* This name is given by Lightfoot in his 'Flora Scotica,' but it should have been *lachan nan damh. Lachan*, a reed, the ox or the hart's reed.

J. squarrosus—Heath-rush, stool-bent. Gaelic: *bru-corcur* (M'Alpine),—*bru-chorachd*, the deers' moor-grass; *bru*, a deer, a hind; *corcach*, a moor or marsh. See *Scirpus.*

> " *Bruchorachd* ás ciob,[1]
> Lusan am bi brigh," &c.
> —M'INTYRE in ' Ben Doran.'
>
> Heath-rush and "deer's hair,"
> Plants nutritious they are, &c.

Specimens of this plant have also been supplied with the Gaelic name *moran* labelled thereon, and in another instance *muran.* These names mean the plants with tapering roots; the same signification in the Welsh, *moron*, a carrot. (See *Muirneach—Ammophila arenaria.*)

J. maritimus and **acutus**—Sea-rush. Irish: *meithan* (O'Reilly). *Meith*, fat, corpulent. *J. acutus* (the great sea-rush) is the largest British species.

Luzula.—Name supposed to have been altered from Italian, *lucciola*, a glow-worm. It was called by the ancient botanists *gramen luxulæ* (Latin, *lux*, light).

L. sylvatica—Wood-rush. Gaelic: *luachar còille*, the bright grass or rush of the wood. The Italian name *lucciola* is said to be given from the sparkling appearance of the heads of flowers when wet with dew or rain. *Learman* (Stewart), possibly from *lear* or *léir*, clear, discernible; a very conspicuous plant, more of the habit of a grass than a rush, the stalk rising to the height of more that two feet, and bearing a terminal cluster of brownish flowers, with large light-yellow anthers.

[1] See *Scirpus cæspitosus.*

85

CYPERACEÆ.

Schœnus (from χοῖνος or σχοῖνος, a cord in Greek).—From plants of this kind cords or ropes were made.

S. nigricans — Bog - rush. Gaelic : *seimhean* (Armstrong). Irish : *seimhin* (*seimh*, smooth, shining—the spikelets being smooth and shining ; or which is more likely, from *siobh* or *siobhag*, straw,—hence *sioman*, a rope made of straw or rushes ; the Greek name σχοῖνος for the same reason).

Scirpus, sometimes written **sirpus** (Freund), seems to be cognate with the Celtic *cirs*, *cors*, a bog-plant ; hence Welsh, *corsfruyn*, a bulrush (Gaelic, *curcais*). Many plants of this genus were likewise formerly used for making ropes. (Cords, Latin, *chorda ;* Welsh, *cord ;* Gaelic and Irish, *corda ;* Spanish, *cuerda*,—all derived from *cors*.)

S. maritimus—Sea-scirpus. Gaelic and Irish : *bròbh*. Name from *brò, brà,* or *bràdh*, a quern, a hand-mill. The roots are large and very nutritious for cattle, and in times of scarcity were ground down in the *muileann bràdh* (French, *moulin à bras*), to make meal ; *bracan*, broth,—hence *bracha*, malt, because prepared by manual labour (Greek, βραχίων ; Latin, *brachium ;* Gaelic, *braic ;* French, *bras*, the arm).

S. cæspitosus—Tufted scirpus, deer's hair, heath club-rush. Gaelic : *ciob, cipe,* and *ciob cheann dubh* (*ciob* = χιβος ; Latin, *cibus*, food ; *ceann*, head ; *dubh*, black).

"Le'n cridheacha' meara
Le bainne na cioba. '—M'INTYRE.

This is the principal food of cattle and sheep in the Highlands in March, and till the end of May. *Cruach luachair,—cruach*, a heap, a pile, a hill, and *luachair*, a rush.

The badge of the Clan M'Kenzie.

S. lacustris—Bulrush, lake-scirpus. Gaelic : *gobhal luachair*, the forked rush (*gobhal*, a fork), from the forked or branched appearance of the cymes appearing from the top of tall, terete (or nearly so), leafless stems. When this tall stem is cut, it goes by the name of *cuilc*,[1] a cane, and is used to bottom chairs. Irish : *gibiun,—gib* or *giob*, rough, and *aoin*, a rush. Gaelic and Irish, *bòg mhuine, boigean, bòg luachair, bòg*,[2] a marsh, a fen, swampy ground, to bob, to wag,—names indicating its habitat,

---

[1] "Mu lochan nan cuilc a tha ruadh."—TIGHMORA.

[2] *Bòg* and *bòlg* are frequently interchanged : *bòlg luachair*, prominent or massy rush ; from *bòlg*, gen. *builg*, comes *bul* in bulrush.

also its top-heavy appearance, causing it to have a bobbing or wagging motion. *Curcais* (*curach*, a marsh, a fen) is more a generic term, and equals *scirpus*. *Min-fheur*, a bulrush. (See *Festuca ovina*.)

Badge of Clan M'Kay.

Eriophorum (from ἔριον, wool, and φέρω, to bear).—Its seeds are covered with a woolly substance—hence it is called cotton-grass.

E. vaginatum and **E. polystachyon**—Cotton-sedge. Scotch: *cat's-tail*. Gaelic and Irish: *canach*. Irish: *cona* (from *can*, white), from its hypogynous bristles forming dense tufts of white cottony down, making the plant very conspicuous in peaty bogs. The *canach* in its purity and whiteness formed the object of comparison in Gaelic poetry for purity, fair complexion, &c., especially in love-songs :—

> "Do chneas mar an *canach*
> Co cheanalta thlà."—M'Intyre.

> Thy skin white as the cotton-grass
> So tender and gentle.

> "Bu ghile na'n *canach* a crùth."—Ossian.

> Her form was fairer than the down of Cana.

In Ossian the plant is also called *caoin cheann* (*caoin*, soft), the soft heads, fair heads.

> "Ghlac mi'n *caoin cheanna* sa' bheinn
> 'Siad ag aomadh mu shruthaibh thall
> Fo charnaibh, bu diomhaire gaoth."—Tighmora.

> I seized cotton-grasses on the hill,
> As they waved by their secret streams,
> In places sheltered from the wind.

This is only the plural form of the name *canach—caineichean*.

> "*Na caineichean* àluinn an t-shléibh."—M'Leod.

O'Reilly gives the name *sgathog fiadhain* to *E. polystachyon*,—*sgath*, a tail, and *og* (dim. termination), the little tail,—to distinguish it from *vaginatum*, which is larger. Scotch: *cat's-tail*.

Badge of Clan Sutherland.

Carex (likely from Welsh, *cors ;* Gaelic, *càrr*, a bog, a marsh, or fenny ground).—This numerous family of plants grows mostly in such situations. *Seisg*, sedge ; *gallsheilisdear*, also *seilisdear amh* (for *Seilisdear*, see *Iris*), — *amh*, raw — the raw sedge. Welsh : *hesg*. *Seasg*, barren, unfruitful. Except *C. rigida*, they are scarcely touched by cattle. According to Dr Hooker, *carex* is derived from Greek, κείρω, from the *cutting* foliage. The Sanscrit root is *kar*, to cut, shear, divide.

C. vulgaris, and many of the other large species—Common sedge. Gaelic: *gainnisg* (Stewart), —*gainne*, a sedge, reed, cane, arrow; and *seasg*.

GRAMINEÆ.

Agrostis alba— Fiorin-grass. Gaelic and Irish: *fioran*, *feorine*, or *fior-than;* derived from Gaelic: *feur*, *feoir*, grass, herbage, fodder. Latin: *vireo*, I grow green,—*ver*, spring; *fœnum*, fodder—*r* and *n* being interchangeable. This name is applied in the dictionaries to the common couch-grass, because, like it, it retains a long time its vital power, and propagates itself by extending its roots.

Alopecurus—Foxtail-grass. Gaelic: *fiteag*,—*fit*, food, refreshment. Latin: *vita*.

A. geniculatus.—Gaelic: *fiteag chàm*,—

> "A chuiseag dheireach's an *fhiteag cham*."—M'INTYRE.

cam, bent, from the knee-like bend in the stalk. A valuable grass for hay and pasture.

Arundo Phragmites—Reed-grass. Gaelic: *seasgan; seasg*, a reed. *Lachan*, the common reed. Irish: *cruisgiornach*, *cruisigh*, music, song; from its stem *reeds* for pipes were manufactured. Welsh: *cawn wellt*, cane-grass; *qwellt*, grass.

Anthoxanthum odoratum — Sweet meadow-grass. Gaelic: *mislean*, from *milis*, sweet.

> "'San canach mìn geal 's *mislean* ann."—M'INTYRE.
>
> The soft white cotton-grass and the sweet grass are there.

Borrach (*borradh*, scent, smell).—In some places this name is given to the *Nardus stricta*, which see. This is the grass that gives the peculiar smell to meadow hay. Though common in meadows, it grows nearly to the top of the Grampians (3400 feet); hence the names are given as "a species of mountain grass" in some dictionaries.

Milium effusum — Millet-grass. Gaelic: *mileid.* Welsh: *miled.* The name derived from the true *millet* misapplied. Millet is translated in the Gaelic Bible *meanbh pheasair*, small peas (see *Faba vulgaris*).—Ezekiel iv. 9.

Phleum pratense—Timothy grass, cat's-tail grass. Gaelic: *bodan*, a little tail; the same name for *Typha angustifolia*. "This grass was introduced from New York and Carolina in 1780 by Timothy Hanson."—LOUDON. It seems to have been unknown in the Hebrides and the Highlands before that date ; for Dr Walker ('Rural Econ. Hebrides,' ii. 27) says, "that it may be introduced into the Highlands with good effect." Yet Lightfoot (1777) mentions it as "by the waysides, and in pastures, but not common." *Bodan* is also applied to *P. arenarium* and *P. alpinum.*

Lepturus filiformis.—Gaelic : *dur fheur fairge*, sea hard grass. *Dur*, hard (Latin, *durus*) ; *feur*, grass ; *fairg*, the sea, ocean, wave. It grows all round Ireland, as well as in England and South Scotland. Irish : *durfher fairge* (O'Reilly).

Calamagrostis.—Etym. κάλαμος, and ἀγρόστις, reed-grass.

C. Epigejos—Wood small reed. *Cuile fheur*, cane-grass ; *gainne* = cane. *Lachan coille*, wood-rush.

Ammophila arenaria (or **Psamma arenaria**)—Sea-maram; sea-matweed. Gaelic and Irish : *muirineach*, from *muir* (Latin *mare*, the sea), the ocean. It is extensively propagated to bind the sand on the sea shore ; generally called *mùran* on west coast. The same name is applied to the carrot, an alteration of *mòron*—a plant with large tapering roots. M'Intyre alludes to "*mùran brìghar*," but whether he refers to the carrot or to this grass is a matter of controversy. Not being a seaside Highlander, he was more likely to know the carrot, wild and cultivated, far better than this seaside grass, and associating it with groundsel (a plant which usually grows rather too abundantly, wherever carrots are sown), makes it a certainty that he had not the "sea-maram" in his mind. (See *Daucus carota.*)

Avena sativa — Oats. Gaelic and Irish : *coirc.* Welsh : *ceirch.* Armoric : *querch.* Probably from the Sanskrit *karç*, to crush.

"Is fheurr slol caol *coirce* fhaotuinn a droch fhearann na' bhi falamh."
Better small oats than nothing out of bad land.

The small variety, *A. nuda*, the naked or hill oat, when ripe, drops the grain from the husk ; it was therefore more generally cultivated two centuries ago. It was made into meal by drying it on the hearth, and bruising it in a stone-mortar, the "*muileann bràdh*"—hand-mill or quern. Many of them may still be seen about Highland and Irish cottages.

A. fatua and **pratensis**—Wild oats. Gaelic : *coirc fiadhain*, wild oats ; *coirc dubh*, black oats. Also applied to the Brome grasses.

" Do'n t-siol chruithneachd, chuireadh gu tiugh;
Cha b' e' n fhiteag, no' n *coire dubh.*"—M'Donald.

When oats become black with blight, the name *coire dubh* is applied, but especially to the variety called *Avena strigosa.*

Hordeum distichon—Barley; the kind which is in common cultivation. ("Barley" comes from Celtic *bàr*, bread, now obsolete in Gaelic, but still retained in Welsh—hence *barn*, and by the change of the vowel, *beer*.) Gaelic and Irish: *eòrna*, *òrna*. Irish: *earn* (perhaps from Latin, *horreo*, to bristle; Gaelic, *òr*, a beard)—O'Reilly. "The bearded or bristly barley;" "*òrog*," a sheaf of corn. *Hordeum*, sometimes written *ordeum* (Freund), is from the same root. "It was cultivated by the Romans for horses, and also for the army; and gladiators in training were fed with it, and hence called *hordiarii.*" It is still used largely in the Highlands for bread, but was formerly made into "crowdie," properly *corrody*, from Low Latin, *corrodium*, a worry.

" Fuarag èorn ann' sàil mo bhroge,
Biadh a b' fhearr a fhuir mi riamh. "

Barley-crowdie in my shoe,
The sweetest food I ever knew.

Irish: *caineog*, oats and barley—from *cain* (Greek, κῆνσος; Latin, *census*), rent, tribute. Rents were frequently paid in "kind," instead of in money.

Secale cereale — Common rye. Gaelic and Irish: *seagall.* Greek: σεχαλη. Armoric: *segal.* French: *seigle.*

" An cruithneach agus an seagall."—Exodus.

The wheat and the rye.

Welsh: *rhyg*, rye.

Molinia cærulea — Purple melic-grass. Gaelic: *bunglàs* (M'Donald), *punglàs.* (*Bun*, a root, a stack; *glàs*, blue.) The fishermen round the west coast and in Skye make ropes for their nets of this grass, which they find by experience will bear the water well without rotting. Irish: *mealoigfèr corcuir* (O'Reilly), —*mealoig*=*melic* (from *mel*, honey), the pith is like honey; *fèr* or *feur*, grass; *corcuir*, crimson or purplish. In some parts of the Highlands the plant is called *braban* (Stewart.)

Glyceria.—From Greek, γλυκύς, sweet, in allusion to the foliage.

G. fluitans—Floating sweet grass. *Milsean uisge, millteach uisge*,—perhaps from *millse*, sweetness. Horses, cattle, and swine are fond of this grass, which only grows in watery places.

M

Trout (*Salmo fario*) eat the seeds greedily. The name *millteach* is frequently applied to grass generally as well as to *Triglochin palustre* (which see). *Feur uisge*, water-grass.

Briza.—Quaking-grass. Gaelic and Irish : *conan,—conan*, a hound, a hero, a rabbit,—may possibly be named after the celebrated "*Conan Maol*," who was known among the Feine for his thoughtless impetuosity. He is called "*Aimlisg na Fèinne*," the mischief of the Fenians. This grass is also called *feur gortach*, hungry, starving grass. "A weakness, the result of sudden hunger, said to come on persons during a long journey or in particular places, in consequence of treading on the fairy grass" —(Irish Superstitions). *Fèur sìthein sìthe*,—literally, a blast of wind; a phantom, a fairy. The oldest authority in which this word *sìthe* occurs is Tirechan's 'Annotations on the Life of St Patrick,' in the Book of Armagh, and is translated "*Dei terreni*," or gods of the earth. *Crìth fhèur*, quaking-grass. *Grigleann* (in Breadalbane), that which is in a cluster, a festoon ; the Gaelic name given to the constellation Pleïades.

Cynosurus.—Etym. κυών, a dog, and οὐρά, a tail.

C. cristatus—Crested dog's-tail. Gaelic : *goinear*, or *goin fheur*, and sometimes *conan* (from *coin*, dogs, and *feur*, grass). Irish : *fèur choinein*, dog's grass.

Festuca.—Gaelic : *féisd*. Irish : *féiste*. Latin : *fastus* and *festus*. French : *feste*, now *fête*. English : *feast*, as applied to grass, good pasture, or food for cattle.

F. ovina—Sheep's fescue - grass. Gaelic and Irish : *feur chaorach*.

"*Mìn-fheur* chaorach."—M'INTYRE.

Soft sheep grass.

This grass has fine sweet foliage, well adapted for feeding sheep and for producing good mutton—hence the name. But Sir H. Davy has proved it to be less nutritious than was formerly supposed. *Mìn-fheur* (Armstrong), is applied to any soft grass —as *Holcus mollis*—to a flag, a bulrush ; as "*mìn-fheur gun uisge*," a bulrush without water (in Job).

Triticum, according to Varro, was so named from the grain being originally ground down. Latin : *tritus*, occurring only in the ablative (*tero*). Greek : τείρω, to rub, bruise, grind.

T. æstivum (and other varieties)—Wheat. Gaelic and Irish : *cruithneachd—cruineachd*. This name seems to be associated with the Cruithne, a tribe or tribes who, according to tradition, came from Lochlan to Erin, and from thence to Alban, where they founded a kingdom which lasted down till the seventh century. Another old name for wheat—*breothan*, may similarly be connected with another ancient tribe, "*Clanna Breogan*. They occupied the territory where Ptolemy in the second century places an offshoot of British Brigantes."—SKENE. Were these tribes so called in consequence of cultivating and using wheat? or was it so called from those tribal names? are questions that are difficult to answer. It seems at least probable that they were among the first cultivators of wheat in Britain and Ireland. *Breothan*, that which is bruised; the same in meaning as *triticum*. Other forms occur, as *brachtan*,[1] being bruised or ground by hand in the "*muileann bràdh*," the quern; sometimes spelled *breachtan*. *Mann*, wheat, food. *Fiormann*, —*fior*, genuine, and *mann*, a name given to a variety called French wheat. *Tuireann*, perhaps from *tuire*, good, excellent. The flour of wheat is universally allowed to make the best bread in the world. *Romhan*, Roman or French wheat; "branks."

T. repens—Couch, twitch. Scotch: *dog-grass*, *quickens*, &c. Gaelic: *feur a'-phuint* (M'Kenzie), the grass with points or articulations. Every joint of the root, however small, having the principle of life in it, and throwing out shoots when left in the ground, causing great annoyance to farmers. (From the root *punc* or *pung*; Latin, *punctum*, a point.) *Goin-fheur*, dogs-grass; or *goin*, a wound, hurt, twitch. According to Rev. Mr Stewart, Nether Lochaber, this name is also given to *Cynosurus*. *Fiothran*, the detestable. It is one of the worst weeds in arable lands on account of the propagating power of the roots. *Bruim fheur*, flatulent grass. Probably only a term of contempt.

T. junceum—Sea-wheat grass. Gaelic: *glàs fheur*, the pale green grass; a seaside grass. It helps, with other species, to bind the sand.

Lolium perenne and **temulentum** — Darnell, rye-grass. Gaelic: *breoillean*. Irish: *breallan* (*breall* or *breallach*, knotty), from the knotty appearances of the spikes, or from its medicinal virtues in curing glandular diseases. "And being used with quick brimstone and vinegar it dissolveth knots and kernels,

---

[1] Latin: *brace* or *brance*. Gallic, of a particularly white kind of corn. According to Hardouin, *blé blanc Dauphiné*, *Triticum Hibernum*, Linn., var. *Granis albis*. Lat., *sandala*.

"Galliæ quoque suum genus farris dedere: quod illie brance vocant apud nos sandalum nitidissimi grani."—PLINY, 18, 7.

and breaketh those that are hard to be dissolved "—CULPEPPER. *Dithean*, darnel ; perhaps from *dìth*, want, poverty. It may be so named from its growing on poor sterile soil, which it is said to improve. " They have lately sown ray-grass to improve cold, clayey soil "—Dr PLATT, 1677. *Roille.* Irish : *raidhleadh*, from *raidhe*, a ray—hence the old English name *ray-grass.* French : *ivraie*, darnel. Welsh : *efr*—perhaps alterations of the French *ivre*, drunk. The seeds of darnel, when mixed with meal, cause intoxication, and are believed to produce vertigo in sheep—the disease that maketh them reel ; and for this reason the grass is often called *sturdan*, from *sturd*,—hence Scotch *sturdy* grass. *Siobhach*, from *siobhas*, rage, fury, madness. " It is a malicious plant of sullen Saturn : as it is not without some vices, so it hath also many virtues "—CULPEPPER. *Cuiseach* (M'Alpine), rye-grass. *Ruintealas* (O'Reilly), the loosening, aperient, or purgative grass—from *ruinnec*, grass, and *tealach*, loosening.

Nardus stricta—Mat - grass, moor - grass. Gaelic : *beitean* (perhaps from *beithe*), was refused. Cattle refuse to eat it. It remains in consequence in dense tufts, till it is scorched by early frosts. In this condition it is frequently burned, in order to destroy it. *Borrach* (in some places), parching. *Carran* (Stewart), a name given also to *Spergula arvensis.* To this grass and other rough species, as rushes, sedges, &c., the name *riasg* is given.

> " Cuiseagan-a's *riasg*
> Chinneas air an t'sliabh."—M'INTYRE.

Aira flexuosa—Waved hair-grass. Gaelic : *mòin-fheur*, peat-grass. It grows generally in peaty soil.

CRYPTOGAMIA.

FILICES.

Filices — Ferns. Gaelic : *raineach*, *roineach.* Irish : *raith*, *raithne*, *raithneach ;* also, *reathnach.* Welsh : *rhedyn.* Perhaps formed from *reath*, a revolution or turning about, or *rat*, motion, from the circinate evolution of the young fronds—an essential characteristic of ferns.

Polypodium vulgare — *Cloch - reathneach* (Armstrong), the stone-fern ; *cloch*, a stone. It is common on stone-walls, stones, and old stems of trees. *Ceis-chrànn.* Irish : *céis chrainn*,— *cis*, a tax, tribute, and *crànn*, a tree, because it draws the substance from the trees ; or from the crosier-like development of the fronds, like a shepherd's crook, " *cis-cean.*" *Sgèamh na cloch.*

Sgeamh means reproach, and *sgiamh* or *sgèimh*, beauty, ornament; "*na cloch*," of the stones. The second idea seems, at least in modern times, to be more appropriate than the first, especially as the term was applied to the really beautiful oak-fern.

> " Mu chinneas luibhean 'us an *sgèimh*."
> How the flowers and the ferns grow.

Reidh raineach,—*reidh*, smooth, plain. *Raineach nan crag*, the rock-fern. *Mearlag* (in Lochaber), perhaps from *meàr* or *meùr*, a finger, from a fancied resemblance of the pinnules to fingers.

P. Dryopteris—Oak-fern. Gaelic and Irish : *sgeamh dharaich* (O'Reilly), the oak-fern. No Gaelic name is recorded for the beech-fern (*P. Phegopteris*).

Blechnum spicant—Hard fern. The only Gaelic name supplied for this fern is "*an raineach chruaidh*," hard fern. It is impossible to say whether this is a translation or not. Being a conspicuous and well-defined fern, it must have had a Gaelic name.

Cystopteris fragilis—Bladder-fern. Gaelic: *friodh raineach*, or *frioth fhraineach,*—"*frioth*," small, slender. The tufts are usually under a foot long ; stalks very slender.

Polystichum aculeatum, lobatum, and **angulare** — Gaelic : *ibhig* (Rev. A. Stewart), the name by which the shield-ferns are known in the West Highlands. This name may have reference to the medicinal drinks formerly made from the powdered roots being taken in water as a specific for worms (see *L. filix-mas*), from *ibh*, a drink. French : *ivre*. Latin : *ebrius*.

P. Lonchitis—Holly fern. Gaelic: *raineach-chuilinn* (Stewart), holly fern, known by that name in Lorne ; also *còlg raineach*, in Breadalbane and elsewhere. For *cuileann* and *còlg*, see *Ilex aquifolium*.

Lastrea Oreopteris—Sweet mountain fern. Gaelic: *crim-raineach* (Stewart). Most likely from *creim*, a scar, the stalks being covered with brown scarious scales. In some places the name *fàile raineach* is given, from *fàile*, a scent, a smell. This species may be easily distinguished by the minute glandular dots on the under side of the fronds, from which a fragrant smell is imparted when the plant is bruised.

L. filix-mas—Male fern. Gaelic and Irish : *marc raineach*, horse-fern. *Marc.* Welsh : *march*. Old High German : *marah*, a horse. This fern has been celebrated from time immemorial as a specific for worms; the powdered roots, taken in water, were considered an excellent remedy. Irish: *raineach-madra*, dog-fern.

L. spinulosa, and the allied species *dilatata* and *Fœnisecii*, are known by the name *raineach nan rodainn*, from Latin, *rodo*. Sanscrit: *rad*, to break up, split, gnaw,—the rat's fern, in Morven, Mull, and Lewis. "Dr Hooker is mistaken as to the range of this fern, as it is extremely abundant here, at least in the form of *dilatata*"—(Lewis Correspondent).[1] The name rat's fern, from its commonness in holes, and the haunts of rats.

Athyrium filix-fœmina—Lady-fern. Gaelic and Irish: *raineach Mhuire*, Mary's fern,—*Muire*, the Virgin Mary, Our Lady; frequently occurring in plant-names in all Christian countries.

Asplenium.— From Greek: α, privative, and σπλην, the spleen.

A. Trichomanes—Black spleenwort. Gaelic and Irish: *dùbh chàsach*, dark-stemmed. *Lus na seilg*, from *sealg*, the spleen. This plant was formerly held to be a sovereign remedy for all diseases of this organ, and to be so powerful as even to destroy it if employed in excess. *Lus a chorrain*. *Urthàlmhan* (O'Reilly),—*ùr*, green, and *talamh*, the earth. As *dùbh chàsach* is the common name for *Trichomanes*—probably *ùr thalmhan* was applied to *A. viride*. *Failtean fionn*, see *A. capillus-Veneris*.

A. Ruta-muraria—Rue fern. Gaelic: *rue bhallaidh*, wall-ruc.

A. Adiantum-nigrum—Gaelic: *an raineach uàine*, the green fern. Irish: *craobh mac fiadh* (O'Reilly),—*craobh*, a tree, a plant, and *muc fiadh*, wild pig or boar.

Scolopendrium vulgare—Hart's-tongue fern. Gaelic: *creamh mac fiadh*, or in Irish, *creamh nam muc fiadh*. Wild boar's wort, a name also given to Asparagus.

Pteris aquilina—Common brake. Gaelic: *an raineach mhòr*, the large fern. *Raith* (see *Polypodium*). The brake is used for various purposes by the Gaels, such as for thatching cottages;

[1] My well-informed correspondent also remarks: "I may mention one or two other plants, regarding which Dr Hooker's information is slightly out. His *Salix repens* is very common here and in Caithness, though absent in at least some parts further south. *Utricularia minor* can easily be found in quantities near the Butt of Lewis; and *Scutellaria minor*, which he allows no further than Dumbarton, grows equally far north, although all I am aware of could be covered by a table-cloth. Another interesting plant, *Eryngium maritimum*, grows in a single sandy bay on our west coast."

and beds were also made of it. It is esteemed a good remedy
for rickets in children, and for curing worms.

Adiantum capillus-Veneris—Maiden-hair fern. Gaelic: *failtean fiònn* (Armstrong), from *falt*, hair, and *fiònn*, fair, resplendent.
This fern is only known in the Highlands by cultivation. This
name is frequently given to *Trichomanes* (*dùbh chàsach*) impro-
perly.

Ophioglossum—From Greek: ὄφις, a serpent, and γλῶσση, a
tongue. The little fertile stalk springing straight out of the
grass may not inaptly be compared to a snake's tongue.

O. vulgatum—Adder's tongue. *Lus na nathraith* (M'Kenzie),
the serpent's weed. *Teanga a' nathrach*, the adder's tongue.
Welsh: *tafad y neidr*, adder's tongue. In the Western Highlands, *beasan* or *feasan* (Stewart).

Osmunda—Osmunder, in Northern mythology, was one of the
sons of Thor (Gaelic: *Tordan*, the thunderer, the Jove of the
Celts; *os* in Celtic, over, above, upon, and *munata*, a champion,
in Irish),—said to have received the name on account of its po-
tential qualities in medicine.

O. regalis—Royal fern. Gaelic: *raineach rioghail*, kingly
fern; *righ raineach*, royal fern. In Ireland it is called bog-
onion.

Botrychium lunaria—Moonwort. Gaelic: *luan lus*, moonwort. Welsh: *y lleñadlys,—lleuad*, moon. "*Luan*, the moon,
seems a contraction of *luath-an*, the swift planet"—ARM-
STRONG. But rather from Sanscrit: *luach*, light. Latin: *luna*.
French: *lune*. *Dèur lus* and *dealt lus* (Stewart),—*dèur*, a tear,
a drop of any fluid, and *dealt*, dew. This plant was held in
superstitious reverence among Celtic and other nations. Horses
were said to lose their shoes where it grew. "On Sliabh Riabhach Mountain no horse can keep its shoes; and to this day it
is said that on Lord Dunsany's Irish property there is a field
where it is supposed all live stock lose their nails if pastured
there." "A Limerick story refers to a man in Clonmel jail who
could open all the locks by means of this plant." The same
old superstition still lingers in the Highlands—

> There is an herb, some say whose virtue's such
> It in the pasture, only with a touch,
> Unshoes the new-shod steed.

"On White-Down, in Devonshire, near Tiverton, there was
found thirty horse-shoes pulled off from the feet of the Earl of
Essex, his horses being there drawn up into a body, many of

them being but newly shod, and no reason known, which caused much admiration ; and the herb described usually grows upon heaths."—CULPEPPER.

Ferns frequently formed components in charms.

> " Faigh naoi gasan rainaich
> Air an gearradh, le tuaigh,
> A's tri chnaimhean seann-duine
> Air an tarruinn à uaigh," &c.—M'INTYRE.

> Get nine branches of ferns
> Cut with an axe,
> And three old man's bones
> Pulled from the grave.

"Fern seeds were looked upon as magical, and must be gathered on Midsummer eve."—Scottish and Irish Superstition.

LYCOPODIACEÆ.

Lycopodium, from λύκος, a wolf, and πούς, a foot, from a fancied resemblance to a wolf's foot.

L. Selago—Fir club-moss. Gaelic : *garbhag an sleibhe,* the rough one of the hill. "The Highlanders make use of this plant instead of alum to fix the colours in dying. They also take an infusion of it as an emetic and cathartic ; but it operates violently, and, unless taken in a small dose, brings on giddiness and convulsions."—LIGHTFOOT. According to De Thèis, "Selago " is derived from the Celtic, *sel (sealladh),* sight, and *jach (loc).* Greek : ἴασις, a remedy, being useful for complaints in the eyes.

Badge of Clan M'Rae.

L. clavatum, annotinum, and the rest of this family are called *lus bhalgaire,* the fox-weed.

EQUISETACEÆ.

Equisetum, from *equus,* a horse, and *seta,* hair, in allusion to the fine hair-like branches of the species. Those plants of this order growing in watery places are called in Gaelic and Irish, *clois, clò-uisge,* the names given to *fluviatile, palustre, ramosum ;* and those flourishing in drier places, *earbuill-each,* horse-tail. *Clois* seems a contraction of *clò-uisge* (O'Reilly),—*clò,* a nail-pen or peg, perhaps suggested by the appearance of the fruiting stems, and *uisge,* water.

E. hyemale—Dutch rushes, shave-grass. Gaelic : *a bhiorag,* —*bior,* a pointed small stick, anything sharp or prickly. Or water (*see* Appendix). This species was at one time extensively used for polishing wood and metal, a quality arising from the cuticle abounding in siliceous cells—hence the use made of the plant for scouring pewter and wooden things in the kitchen. A large quantity used to be imported from Holland, hence the name " Dutch rushes." Irish : *gadhar*, from *gad*, a withe, a twig. *Liobhag*, from *liobh*, smooth, polish. It grows in marshy places and standing water. *Cuiridin* (O'Reilly), because growing on marshy ground.

BRYACEÆ.

Gaelic and Irish : *coinneach, caoineach*, from *caoin*, soft, lowly, &c. The principal economic use of moss to the ancient Gaels was in making bed-stuffs, just as the Laplanders use it to this day.

" Trì coilceadha na Feinne, bàrr gheal chrann, *coinneach*, 'us ùr luachair."

The three Fenian bed-stuffs—fresh tree-tops, *moss*, and fresh rushes.

Welsh : *mwswg*, moss.

Sphagnum—Bog-moss. Gaelic: *mointeach liath* (*moin*, peat, and *liath*, grey). From its roots and decayed stalks peat is formed. *Fionnlach*, from *fionn*, white. It covers wide patches of bog, and when full grown it is sometimes almost white ; occasionally the plant has a reddish hue (*coinneach dhearg*, red moss). Martin refers to it in his ' Western Islands : ' " When they are in any way fatigued by travel or otherways, they fail not to bathe their feet in warm water wherein *red moss* has been boiled, and rub them with it on going to bed." This seems to be the only moss having a specific name in Gaelic, the rest going by the generic term *còinneach*.

" Còinich uine mu 'n iomall,
A's imadach seòrsa."—M'INTYRE.

Green moss around the edges,
Many are the kinds.

MARCHANTIACEÆ AND LICHENES.

Marchantia polymorpha—Liverwort. Gaelic: *lus an àinean*, the liverwort. Irish : *cùisle aibheach*. Welsh : *llysiar afu—afu*, the liver. (Names derived from its medicinal effects on the liver.) Irish : *duilleog na crùithneachta*, the leaf of (many) shapes or forms. *Crùth*, form, shape, synonymous with Greek "*polymorpha*."

Peltidea canina—The dog-lichen. Gaelic: *lus ghonaich* (from *gòin*, wound ; *gòineach*, agonising). This plant was formerly used for curing distemper and hydrophobia in dogs. The name "*gearan*, the herb dog's-ear," is given in the dictionaries.

Probably this name was applied to this plant, meaning a complaint, a groan. Welsh: *gerain*, to squeak, to cry.

Lecanora.—Etymology of this word uncertain (in Celtic, *lech* or *leac*, means a stone, a flag). Greek: λίθος.

L. tartarea—Cudbear. Gaelic and Irish: *corcar* or *corcur*, meaning purple, crimson. This lichen was extensively used to dye purple and crimson. It is first dried in the sun, then pulverised and steeped, commonly in urine, and the vessel made air-tight. In this state it is suffered to remain for three weeks, when it is fit to be boiled in the yarn which it is to colour. In many Highland districts many of the peasants get their living by scraping off this lichen with an iron hoop, and sending it to the Glasgow market. M'Codrum alludes to the value of this and the next lichen in his line

> "Spréigh air mointich,
> Or air chlachan."
> Cattle on the hills,
> Gold on the stones.

Parmelia saxatilis and **omphalodes**—Stone and heath parmelia. Gaelic and Irish: *crotal*. These lichens are much used in the Highlands for dyeing a reddish brown colour, prepared like *tartarea*. And so much did the Highlanders believe in the virtues of *crotal* that, when they were to start on a journey, they sprinkled it on their hose, as they thought it saved their feet from getting inflamed during the journey. Welsh: *cen dû*, black head, applied to the species *Omphalodes*.

Sticta pulmonacea (*Pulmonaria* of Lightfoot) — Lungwort lichen. Scotch: *hazelraw*. Gaelic and Irish: *crotal coille* ("*coille*" of the wood), upon the trunks of trees in shady woods. It was used among Celtic tribes as a cure for lung diseases, and is still used by Highland old women in their ointments and potions.

According to Shaw, the term *grim* was applied as a general term for lichens growing on stones. Martin, in his description of his journey to Skye, refers to the superstition "that the natives observe the decrease of the moon for scraping the scurf from the stones." The two useful lichens, *corcur* and *crotal*, gave rise to the suggestive proverb—

"Is fhearr a' chlach gharbh air am faighear rud-eigin, na 'chlach mhìn air nach faighear dad idir."

Better the rough stone that yields something, than the smooth stone that yields nothing.

FUNGI.

Agaricus—The mushroom. Irish and Gaelic dictionaries give *agairg* for mushroom. Welsh : *cullod*.

A. campestris—*Balg bhuachail* (*balg* is an ancient Celtic word, and in most languages has the same signification—viz., a bag, wallet, pock, &c. (Greek, βολγυς; Latin, *bulga ;* Sax. *belge ;* Ger. *bálg*), *buachail*, a shepherd). *Balg losgainn* (*losgann* a frog, and in some places *bàlg bhuachair,—buachar*, dung), *Leirin sugach*. In Aberfeldy *A. campestris* is called *bonaid bhuidhli smachain* (Dr M'Millan).

Boletus bovinus—Brown boletus. Gaelic and Irish : *bonaid an losgainn*, the toad's bonnet ; and also applied to other species of this genus.

Tuber cibarium—Truffle. *Ballan losgainn*, Dr M'Millan, from *ball*, a ball, a tuber. These are subterraneous ball-like bodies, something like potatoes, found in beech-woods in Glen Lyon ; and probably applied to other species as well.

Lycoperdon giganteum—The large fuz-ball or devil's snuff-box. Gaelic and Irish : *beac, beacan*, from *beach*, a bee. This mushroom or puff-ball was used formerly (and is yet) for smothering bees ; it grows to a large size, sometimes even two or three feet in circumference. *Trioman* (O'Reilly).

L. gemmatum—The puff-ball, fuz-ball. Gaelic and Irish: *caochag*, from *caoch* (Latin, *cæcus*), blind, empty, blasting. It is a common idea that its dusty spores cause blindness. *Bàlg smùid*, the smoke-bag ; *bàlg séididh*, the puff-bag. *Bàlg peiteach bocan*, or *bochdan-bearrach* (*bochdan*, a hobgoblin, a sprite, and *bearr*, brief, short), and *bonaid an losgainn*, are frequently applied to all the mushrooms, puff-balls, and the whole family of the larger fungi.

Polyporus.—The various forms of cork-like fungi growing on trees are called *caise* (Irish), meaning cheese, and in Gaelic *spuing* or (Irish) *spuinc*, sponge, from their porous spongy character.

P. fomentarius and **betulinus**—Soft tinder. Gaelic : *cailleach spuinge*, the spongy old woman,—a corruption of the Irish *caisleach spuine*, soft, cheese-like sponge. It is much used still by Highland shepherds for making *amadou* or tinder, and for sharpening razors.

Mucedo—Moulds. Gaelic : *cloimh liath*, grey down. Mildew, *milcheo*.

Mushrooms bear a conspicuous part in Celtic mythology from their connection with the fairies,—they formed the tables for their merry feasts. Fairy rings (*Marasmius oreades*, other species of *Agarici*) were unaccountable to our Celtic ancestors save by the agency of supernatural beings.

ALGÆ.

The generic names assigned to sea-weeds in Gaelic are : *feamainn* (*feam*, a tail) ; *trailleach* (M'Alpine), (from *tràigh*, shore, sands) ; *barra-rochd* (*barr*, a crop), *roc.* Greek : ῥώξ. French : *roche*, a rock. Welsh : *gwymon*, sea-weed. French : *varec*, from Sanscrit, *bharc*, through the Danish *vrag*. All the olive-coloured sea-weeds go by the general name *feamainn buidhe ;* the dark-green, *feamainn dubh ;* and the red, *feamainn derg.*

Fucus vesiculosus — Sea-ware, kelp-ware, black tang, lady-wrack. Gaelic : *propach*, sometimes *prablach*, tangled ; in some places *gròbach, gròb*, to dig, to grub.

This fucus forms a considerable part of the winter supply of food for cattle, sheep, and deer. In the Hebrides cheeses are dried without salt, but are covered with the ashes of this plant, which abounds in salt. It was also used as a medicinal charm. " If, after a fever, one chanced to be taken ill of a stitch, they (the inhabitants of Jura) take a quantity of *lady-wrack* and *red fog* and boil them in water ; the patients sit upon the vessel and receive the fume, which by experience they find effectual against the distemper." — MARTIN's ' Western Isles.'

F. nodosus—Knobbed sea-weed. Gaelic: *feamainn bholgainn, builgach,—bolg, builg*, a sack, a bag, from the vesicles that serve to buoy up the plant amidst the waves. *Feamainn buidhe*, the yellow wrack. It is of an olive-green colour; the receptacles are yellow.

F. serratus—Serrated sea-weed. Gaelic: *feamainn dubh*, black wrack. *Aon chasach*, one-stemmed, applies to this plant when single in growth.

F. canaliculatus—Channelled fucus. Gaelic : *feamainn chir-ean* (*cìr*, a comb). This plant is a favourite food for cattle, and farmers give it to counteract the injurious effects of sapless food, such as old straw and hay.

Laminaria digitata—Sea-girdles, tangle. Gaelic and Irish : *stamh, slàt-mhàra*, sea-wand. *Duidhean*, the stem, and *liaghag* or *leathagan*, *bàrr stamh*, and *bragair*, names given to the broad leaves on the top. *Doire* (in Skye), tangle. Though not so much used for food as formerly, it is still chewed by the Highlanders when tobacco becomes scarce. It was thought to be an effectual remedy against scorbutic and glandular diseases, even long before it was known to contain iodine. " A rod about four, six, or eight feet long, having at the end a blade slit into seven or eight pieces, and about a foot and a half long. I had an account of a young man who lost his appetite and had taken pills to no purpose, and being advised to boil the blade of the Alga, and drink the infusion boiled with butter, was restored to his former state of health "—MARTIN's ' Western Isles.' By far the most important use to which this plant and the other fuci have been put was the formation of kelp; much employment and profit were derived from its manufacture : *e.g.*, in 1812, in the island of North Uist, the clear profits from the proceeds of kelp amounted to £14,000; but the alteration of the law regarding the duty on barilla reduced the value to almost a profitless remuneration of only £3500.

L. saccharina—Sweet tangle, sea-belt. Gaelic : *smeartan* (*smear*, greasy). The Rev. Mr M'Phail gives this name to " one of the red sea-weeds." Other correspondents give it to this plant.

L. bulbosa—Sea furbelows, bulbous-rooted tangle. Gaelic : *sgrothach*. This name is doubtful (*sgroth*, pimples, postules).

Alaria esculenta—Badderlocks, hen-ware (which may be a contraction of honey-ware, the name by which it is known in the Orkney Islands). Gaelic : *mircean* (one correspondent gives this name to " a red sea-weed "), seemingly the same as the Norse name *Mária kjerne*,—*Mári*, Mary, and *kjerne* is our word kernel, and has a like meaning. In Gaelic and Irish dictionaries, *muirirean* (Armstrong), *muiririn* (O'Reilly), " a species of edible alga, with long stalks and long narrow leaves "—SHAW. In some parts of Ireland, Dr Drummond says, it is called *murlins*—probably a corruption of *muiririn*, *muirichlinn*, *muirlinn* (M'Alpine), (from *muir*, *mara*, the sea). It is known in some parts of Ireland by the name *sparain* or *sporain*, purses, because the pinnated leaflets are thought to resemble the Highlander's *sporan*. *Gruaigean* (in Skye).

Rhodymenia palmata—Dulse. Gaelic and Irish : *duiliasg*, from *duille*, a leaf, and *uisge*, water—the water-leaf. The Highlanders and Irish still use *duiliasg*, and consider it wholesome when eaten fresh. Before tobacco became common, they used to prepare dulse by first washing it in fresh water, then drying it in the sun : it was then rolled up fit for chewing. It was also used medicinally to promote perspiration. *Fithreach*, dulse. *Duiliasg staimhe* (*staimh, Laminaria digitata*). It grows frequently on the stems of that fucus. *Duiliasg chlaiche—i.e.*, on the stones, the stone dulse. *Duileasg* is also given to *Laurentia pinnatifida*, formerly eaten under the name of pepper dulse.

Porphyra laciniata—Laver, sloke. Gaelic and Irish : *sloucan, slochdan,* from *sloc*, a pool or slake. *Slàbhcean* (in Lewis), *slàbhagan* (Shaw). Lightfoot mentions that "the inhabitants of the Western Islands gather it in the month of March, and after pounding and stewing it with a little water, eat it with pepper, vinegar, and butter ; others stew it with leeks and onions.

Ulva latissima—Green ulva. Gaelic : *glasag*, also applied to other edible sea-weeds. In some places in the Western Highlands the names given to laver are also given to this plant. *Glasag*, from *glàs*, blue, or green.

Palmella montana (Ag.)—Lightfoot describes, in his ' Flora Scotica,' a plant which he calls *Ulva montana*, and gives it the Gaelic name *duileasg nam beann—i.e.*, the mountain dulse. This plant is *Gloeocapsa magma* (Kutzing). *Protococcus magma* (Brebisson, Alg. Fallais). *Sorospora montana* (Hassall). Lightfoot was doubtless indebted to Martin (whose ' Western Isles ' furnished him with many of his useful notes on the uses of plants among the Highlanders) for the information respecting such a plant. Martin describes it thus : " There is seen about the houses of Bernera, for the space of a mile, a soft substance resembling the sea-plant called *slake* [meaning here *Ulva latissima*], and grows very thick among the grass ; the natives say it is the product of a dry hot soil ; it grows likewise *on the tops of several hills* in the island of Harris." " It abounds in all mountainous regions as a spreading crustaceous thing on damp rocks, usually blackish-looking ; but where it is thin the purplish nucleus shines through, giving it a brighter aspect."—Roy.

Chondrus crispus—Irish moss, known in the Western Highlands by the Irish name *an carraceen*, as the chief supply used to come from Carrageen in Ireland. At one time it was in much repute, for from it was manufactured a gelatinous easily digested food for invalids, which used to sell for 2s. 6d. per lb. *Mathair* *an duileasg*, the mother of the dulse, as if the dulse had sprung from it.

Corallina officinalis.—Gaelic: *coireall* (M'Alpine). Latin: *corallium*, coral. *Linean*. It was used as a vermifuge.

Polysiphonia fastigiata. A tuft of this sea-weed was sent to me with the Gaelic name *Fraoch màra*, sea-heather, written thereon.

Hemanthalia lorea.—The cup-shaped frond from which the long thongs spring is called *aiomlach*, or *iomleach* (*iomleag*, the navel), from the resemblance of the cup-shaped disc to the navel. Dr Neill mentions that in the north of Scotland a kind of sauce for fish or fowl, resembling ketchup, is made from the cup-like or fungus-like fronds of this sea-weed.

Halydris siliquosa.—Gaelic: *roineach mhàra*, the sea-fern. (In the Isle of Skye.)

Chorda filum—Sea-laces. In Shetland, Lucky Minny's lines; Ayrshire, dead men's ropes. Gaelic: *gille mu leann* (or *mu lìon*),—*gille*, a young man, a servant; *lìon*, a net. Lightfoot mentions that the stalks acquire such toughness as to be used for fishing lines, and they were probably also used in the manufacture of nets. At all events it is a great obstacle when trawling with nets, as it forms extensive sea-meadows of long cords floating in every direction. In some parts *langadair* is given to a "sea-weed, by far the longest one." This one is frequently from twenty to forty feet in length.

Sargassum vulgare (or **bacciferum**)—Sea-grapes. Gaelic: *tùrusgar* (sometimes written *trusgar*, from *trus*, gather), from *tùrus*, a journey. This weed is frequently washed by the Gulf Stream across the great Atlantic, with beans, nuts, and seeds, and cast upon the western shores. These are carefully gathered, preserved, and often worn as charms. They are called *uibhean sìthein*, fairy eggs, and it is believed that they will ward off evil-disposed fairies. The nuts are called *cnothan-spuinge*, and most frequently are *Dolichas urens* and *Mimosa scandens*. To *Callithamnion Plocamium*, &c., and various small red sea-weeds, such as adorn ladies' albums, the Gaelic name *smòcan* is applied.

Confervæ, such as *Enteromorpha* and *Cladophora*. Gaelic and Irish: *lianach* or *linnearach* (*linne*, a pool). Martin describes a plant under the name of *linarich*—"a very thin, small, green plant, about eight, ten, or twelve inches in length; it grows on stones, shells, and on the bare sands. This plant is applied plasterwise to the forehead and temples to procure sleep for

such as have a fever, and they say it is effectual for the purpose."
—MARTIN'S ' Hebrides.' *Barraig uaine,* the green scum on
stagnant water. *Feuruisge,* water-grass. *Lochan. Griobhars-
gaich,* the green scum on water.

" Tha uisge srùth na dìge
    Na shrùthladh dùbh gun sioladh
Le barraig uaine, liògh ghlas,
    Gu mi bhlasda grannd,
Fèur lochan is *tachair*
    An cinn an duileag bhàite."—M'INTYRE.

The water in its channel flows,
    A dirty stagnant stream,
And algæ green, like filthy cream,
    Its surface only shows.
With water-grass, a choking mass,
    The water-lily grows.

APPENDIX.

ADDITIONAL GAELIC NAMES.

These names were either unintentionally omitted, or did not come under my observation until too late for insertion in their proper botanical order.

Airgiod luachra (*Spirea ulmaria*)—Meadow-sweet, meaning the silvery rush. *Airgiod.* Latin : *argentum.*

Amharag (*Sinapis arvensis*)—Cherlock. From the root *amh*, raw or pungent, and probably corrupted into *"Marag" bhuidhe* (page 7); also in *Cochlearia officinalis. A'maraich* (page 5), for *amharaich*, from the same root, on account of the pungent taste of both plants.

Barr a-bhrigean (*Potentilla anserina*)—Silverweed.

Bath ros (*Rosmarinus officinalis*)—Rosemary. From *bàth*, the sea ; and *ròs*, a rose.

Bearnan bearnach (*Taraxacum dens-leonis*)—Dandelion.

Bearnan bealtine (*Caltha palustris*)—Marsh-marigold.

Billeog an spuinc (*Tussilago farfara*)—Coltsfoot (page 41).

Biodh an 't sionaidh (*Sedum anglicum*). (*Sionaidh*, a prince, a lord, chief; *biodh*, food.) From the name it is evident that the plant was formerly eaten, and considered a delicacy.

Bior ros (*Nymphæa*)—Water-lily. *Bior*, or its aspirated form *bhir* or *bhior*, meaning water; in Arabic, *bir ;* Hebrew, *beer.* From this root comes the name *bhiorag*, a water-plant (*Equisetum hyemale*, page 96), and such place and river names as *ver* in Inver, *her* in Hereford, and the river *Wear* in Durham.

Blath nam bodaigh (*Papaver*)—Poppy, meaning the rustic's flower.

Bo-coinneal (*Erysimum alliaria*)—Sauce alone. *Bò*, a cow ; *coinneal*, a candle.

Buidhechan-bo-bleacht (*Primula veris*)—Cowslip. The milk-cow's daisies (page 57).

Cal Phadruigh (*Saxifraga umbrosa*)—London pride ; Peter's kale.

Cannach (*Myrica gale*)—Bog-myrtle. (This name must not be

o

confounded with *canach*, the bog-cotton.) It means any fragrant shrub, pretty, beautiful, mild, soft.

Caorag leana (*Lychnis flos-cuculi*)—Ragged robin. *Caorag*, a spark ; and *leana*, a marsh.

Caor con (*Viburnum opulus*)—Dogberry. *Caor*, a berry ; *còn*, dog.

Cerrucan (*Daucus carota* and *Sium sisarum*)—Skirrets. Name applied to the roots of these and the next plant.

Curran earraich (*Potentilla anserina*) — Silver-weed ; wild skirret (page 19).

> " Mil fo thalamh, currain Earraich."
>
> Underground honey, spring carrots,

"Exceptional luxuries. The spring carrot is the root of the silver-weed."—Sheriff NICOLSON.

Coirean coilleach (*Lychnis diurna*) (page 8).

Collaidin ban (*Papaver*)—White poppy (page 4).

Corran lin (*Spergula arvensis*)—Spurrey.

Cuirinin (*Nymphæa*)—Water-lily.

Daileag (*Phœnix dactylifera*)—The date-tree.

Dearag thalmhainn (*Fumaria officinalis*)—Fumitory. From *dearg*, red ; *thalamh*, earth, ground.

Dearcan dubh (*Ribes nigrum*)—Black currants. For *dearc* see page 45.

Deochdan dearg (*Trifolium pratense*)—Red clover.

Driuch na muine (*Drosera rotundifolia*)—Sun-dew. *Driuch*, dew ; and *na muine* of the hill.

Dun, lus (*Scrophularia nodosa*) — Figwort, the high plant. According to Bede *dùn* means a height in the ancient British language ; hence the terminations of names of towns, *don* and *ton*.

Eabh (*Populus tremula*)—Aspen. The Gaelic for Eve.

Eanach (*Nardus stricta*).

Easdradh (*Filices*)—Ferns.

Eidheann mu chrann (*Hedera helix*)—The ivy (page 28).

> " Gach fiodh 's a' choille
> Ach *eidheann mu chrann* a's fiodhagach."
>
> Every tree in the wood
> Except ivy and bird-cherry tree.

Feathlog fa chrann (*Lonicera periclymenum*) (page 34).

Fib (*Vaccinium vitis idæa*)—Whortleberry.

Fineal ghreugach (*Trigonella*)—Greek fennel.

Fiodh almug (*Santalum album*)—Sandal-wood.

"Agus mar an ceudna loingneas Hiram a ghiulain òr o Ophir, agus ro mhoran do *fhiodh almuig*."—(STUART) 1 Kings x. 11.
The navy of Hiram brought in from Ophir gold and great plenty of Almug trees.

Fionnach (*Nardus stricta*)—From *fionn*, white.

Fiuran and **giuran** (*Heracleum spondylium*)—Cow-parsnip.

Fofannan min (*Sonchus oleraceus*)—Sow-thistle. For *fofannan*, see *fothannan* (page 38).

Forr dris (*Rubus rubiginosa*)—Sweet-briar.

Fuaim an t' Siorraigh (*Fumaria officinalis*)—Fumitory. *Fuaim*, sound ; *an t' Siorraigh*, of the sheriff! Probably only a humorous play on the words "*fumaria officinalis.*"

Furran (*Quercus robur*)—The oak.

Gairleach collaid (*Erysimum alliaria*)—Jack by the hedge ; meaning hedge garlic.

Gairteog (*Pyrus malus*)—Crab-apple. From *garg*, sour, bitter.

Gall pheasair (*Lupinus*)—Lupin (see page 16).

Gall uinnseann (*Pyrus aria*)—Quickbeam tree.

Gearr bochdan (*Cakile maritima*)—Sea gilly-flower.

Glaodhran (*Oxalis* and *Rhinanthus crista-galli*)—Meaning a "rattle." Dictionaries give this name to wood-sorrel ; in Breadalbane it is applied generally to the yellow rattle.

Glocan (*Prunus padus*)—Bird-cherry. *Glocan* or *glacan*, a prong or fork.

Goirgin garaidh (*Allium ursinum*)—Garlic.

Goirmin searradh (*Viola tricolor*)—Pansy ; heart's-ease.

Gran arcain (*Ranunculus ficaria*)—Lesser celandine. *Arc*, a cork, from its cork-like roots.

Leamhnach (*Potentilla tormentilla*) — Common tormentil. Name in Gaelic, meaning "tormenting," from which "*leannartach*" probably is a corruption (see page 19).

Leacan, Loan cat } (*Cotyledon umbilicus*)—Navel-wort.

Lochal mothair (*Veronica beccabunga*)—Brook-lime.

Lusra na geire-boirnigh (*Arbutus uva-ursi*)—Red bear berry, the plant of bitterness. *Geire*, bitterness ; and *boirnigh*, feminine. See "*meacan easa fiorine.*"

Lus na meala mor (*Malva sylvestris*)—The common mallow.

Lus mor. Also applied to *Verbascum thapsus*, Mullein, as well as to the foxglove (*Digitalis*).

Lus ros (*Geranium Robertianum*)—Herb Robert ; crane's-bill : the rose-wort.

Lus an lonaidh (*Angelica sylvestris*)—Wood angelica. *Lonaidh* is the piston or handle of the churn. The umbelliferous flower has much the appearance of that implement. The common name in Breadalbane (see page 31).

Lus an t' seann duine—The old man's plant. Name given in some places to "southernwood," *Artemisia abrotanum*.

Lus na seabhag—Hawkweed.

Meacan easa beanine (*Pæonia*)—Female pæony.

Meacan easa fiorine (*Pæonia*)—Male pæony. Old botanists used to distinguish between two varieties of this plant, and named them male and female. This was a mere fanciful distinction, and had no reference to the real functions of the stamens and pistils of plants; but yet there existed a vague idea, from time immemorial, that fecundation was in some degree analogous to sexual relationship, as in animals — hence such allusions as " *Tarbh, coille,*" " *Dair na coille* " (see page 68).

Meilise (*Sisymbrium officinale*)—Hedge mustard.

Neandog chaoch (*Lamium*)—Dead nettle ; blind nettle.

Onn. Some authorities give this name to *Ulex europæa*, as well as to *Euonymus*. Welsh, *chwyn*—hence Scotch and English *whin*.

Pea·air tuilbh (*Orobus tuberosus*)—Bitter vetch.

Ponair churraigh (*Menyanthes*)—Marsh trefoil, meaning the marsh-bean, bog-bean.

Pis phreachain (*Vicia sativa*)—Pis = peas. *Preachan*, a ravenous bird.

Raibhe (*Raphanus*)—Radish.

Ramasg—Applied to various species of *Fuci*, from *ram*, a branch, an oar = oar-weed.

Reagha maighe,
Reagaim and **raema** } (*Sanicula europæus*)—Wood-sanicle.

Reilige, reilteag (*Geranium Robertianum*)—From *reil* or *reul*, a star.

Rian roighe (*Geranium Robertianum*)—Crane's-bill.

Ros mall (*Althæa rosea*)—Hollyhock.

Rotheach tragha (*Crambe maritima*)—Seakale.

Searbhan muic (*Cichorium endiva*)—Endive.

Seircean mor (*Arctium lappa*)—Burdock.

Seud (*Hypericum*).

Sibhin (*Scirpus lacustris*)—Bulrush.

Siode, lus (*Lychnis flos-cuculi*)—Ragged Robin ; meaning the silk-weed, from its silken petals.

Son duileag (*Lapsana communis*)—Nipple-wort. *Son*, good ; *duileag*, a leaf.

Spɔg na cubhaig (*Viola tricolor*)— Pansy, heart's-ease ; meaning the cuckoo's claw.

Spriunan (*Ribes nigrum* and *rubrum*)—Currants.

Straif (*Prunus spinosa*)—Sloe.

Sreang thrian (*Ononis arvensis*)—Rest-harrow.

Staoin (*Nepeta glechoma*)—Also applied to ground - ivy in some places, as well as to juniper.

Subh nam ban sithe (*Rubus saxatilis*)—Stone-bramble ; the fairy-woman's strawberry.

Toir-pin (*Sempervivum tectorum*)—House-leek ; probably the same as *tir-pin* (see page 27).

Traithnin (*Geum urbanum*)—Geum.

Treabhach (*Barbarea vulgaris*)—Winter cress. *Treabh*, a tribe, a village.

Truim crann (*Sambucus niger*)—Elder, corruption from *drum* (see page 34).

Tuile thalmhainn (*Ranunculus bulbosus*)— *Tuile*, a water-course.

Tuimpe—Turnip.

NOTES.

Page 6.

Nasturtium officinalis—Water-cress. A curious old superstition respecting the power of this plant as a charm to facilitate milk-stealing was common in Scotland and Ireland. " Not long ago, an old woman was found, on a May morning, at a spring-well cutting the tops of water-cresses with a pair of scissors, muttering strange words, and the names of certain persons who had cows, also the words, " S' liomsa leath do choud sa" (half thine is mine). She repeated these words as often as she cut a sprig, which personated the individual she intended to rob of his milk and cream." "Some women make use of the root of groundsel as an amulet against such charms, by putting it amongst the cream."— MARTIN. Among the poorer classes, water-cress formed a most important auxiliary to their ordinary food. "If they found a plot of water-cresses or Shamrock, there they flocked as to a feast for the time."—SPENCER.

Page 8.

Drosera rotundifolia—Sun-dew. *Lus na fearnaich.* "*Earnach*" was the name given to a distemper among cattle, caused, it is supposed, by eating a poisonous herb. Some say the sun-dew—others, again, aver the sun-dew was an effectual remedy. This plant was much employed among Celtic tribes for dyeing the hair.

Page 8.

Saponaria. The quotation from Pliny may be thus translated: "Soap is good—that invention of the Gauls—for reddening the hair, out of grease and ash."

Page 9.

Linum usitatissimum (*Lìon*).

> "Mèirle salainn 's mèirle frois,
> Mèirl' o nach fhaigh anam clos ;
> Gus an teid an t-iasg air tìr,
> Cha 'n fhaigh mèirleach an lìn clos."

"This illustrates the great value attached to salt and lint, especially among a fishing population, at a time when the duty on salt was excessive, and lint was cultivated in the Hebrides."—Sheriff NICOLSON.

Page 10.

Hypericum. Martin evidently refers to this plant, and calls it "*Fuga dæmonum.*" "John Morrison, who lives in Bernera (Harris), wears the plant called "*.Seud*" in the neck of his coat to prevent his seeing of visions, and says he never saw any since he first carried that plant about with him." Children have a saying when they meet this plant—

> "Luibh Cholum Chille, gun sireadh gun iarraidh,
> 'Sa dheòin Dia, cha bhàsaich mi nochd."

St Columbus-wort, unsought, unasked, and, please God, I won't die to-night.

Page 12.

Shamrock —Wood-sorrel and white clover. The shamrock is said to be worn by the Irish upon the anniversary of St Patrick for the following reason : When the Saint preached the Gospel to the pagan Irish, he illustrated the doctrine of the Trinity by showing them a trefoil, which was ever afterwards worn upon the Saint's anniversary. "Between May-day and harvest, butter, new cheese, and curds and shamrock, are the food of the meaner sort all this season."—PIERS's 'West Meath.'

Page 13.

Gaelic Alphabet. Antecedent to the use of the present alphabet, the ancient Celts wrote on the barks of trees. The writing on the bark of trees they called *oghuim*, and sometimes trees, *feadha*, and the present alphabet *litri* or letters.

" Cormac Casil cona churu,
Leir Mumu, cor mela :
Tragaid im righ Ratha Bicli,
Na *Litri* is na *Feadha*."

Cormac of Cashel with his companions
Munster is his, may he long enjoy ;
Around the King of Raith Bicli are cultivated
The LETTERS and the TREES.

The "letters" here signify, of course, our present Gaelic alphabet and writings ; but the "trees" can only signify the *oghuim*, letters, which were named after trees indigenous to the country."—Prof. O'CURRY.

Page 16.

Orobus tuberosus (*Corra meille*, M'Alpin, and *cairmeal*, Armstrong) — Bitter vetch — and sometimes called "wild liquorice"—seems to be the same name as the French "*caramel*," burnt sugar ; and according to Webster, Latin, "*canna mellis*," or sugar-cane. The fermented liquor that was formerly made from it, called *cairm* or *cuirm*, seems to be the same as the "*courmi*" which Dioscorides says the old Britons drank. The root was pounded and infused, and yeast added. It was either drunk by itself, or mixed with their ale—a liquor held in high estimation before the days of whisky ; hence, the word "*cuirm*" signifies a feast. That their drinking gatherings cannot have had the demoralising tendencies which might be expected, is evident, as they were taken as typical of spiritual communion. In the Litany of "Aengus Céilé Dé," dating about the year 798, we have a poem ascribed to St Brigid, now preserved in the Burgundian Library, Brussels.

" Ropadh maith lem corm-lina mor,
Do righ na righ,
Ropadh maith lem muinnter nimhe
Acca hol tre bithe shir."

I should like a great lake of ale
For the King of kings ;
I should like the family of heaven
To be drinking it through time eternal.

To prevent the inebriating effects of ale, "the natives of Mull are very careful to chew a piece of "*charmel*" root, finding it to be aromatic—especially when they intend to have a drinking-

bout ; for they say this in some measure prevents drunkenness."
—MARTIN's 'Western Isles.'

Trees, Thorns. A superstition was common among the Celtic races, that for every tree cut down in any district, one of the inhabitants in that district would die that year. Many ancient forts, and the thorns which surrounded them, were preserved by the veneration, or rather dread, with which the thorns were held ; hence, perhaps, the name *sgitheach, sgith* (anciently), fear ; hence also, *droighionn (druidh)*, enchantment, witchcraft.

Page 20.

Rubus fruticosus —(*Smearagan*) Blackberries. It was and is, I believe, still a common belief in the Highlands that each blackberry contains a poisonous worm. Another popular belief is—kept up probably to prevent children eating them when unripe—that the fairies defiled them at Michaelmas and Halloween.

Page 24.

Pyrus aucuparia—(*Craobh chaoran*) Mountain-ash. The Highlanders have long believed that good or bad luck is connected with various trees. The *caoran* or *fuinnseach coille* (the wood enchantress) was considered by them as the most propitious of trees ; hence, it was planted near every dwelling-house, and even far up in the mountain-glens, still marking the spot of the old shielings. "And in fishing-boats as are rigged with sails, a piece of the tree was fastened to the haul-yard, and held as an indispensable necessity." "Cattle diseases were supposed to have been induced by fairies, or by witchcraft. It is a common belief to bind unto a cow's tail a small piece of mountain-ash, as a charm against witchcraft."—MARTIN. And when malt did not yield its due proportion of spirits, this was a sovereign remedy. In addition to its other virtues, its fruit was supposed to cause longevity. In the Dean of Lismore's Book there occurs a very old poem, ascribed to Caoch O'Cluain (Blind O'Cloan) ; he described the rowan-tree thus—

> "Caorthainn do bhi air Loch Maoibh do chimid an traigh do dheas,
> Gach a ré 'us gach a mios toradh abuich do bhi air.
> Seasamh bha an caora sin, fa millise no mil a bhlàth,
> Do chumadh a caoran dearg fear gun bhiadh gu ceann naoi tràth,
> Bleadhna air shaoghal gach fir do chuir sin is sgeul dearbh."

> A rowan-tree stood on Loch Mai,
> We see its shore there to the south ;
> Every quarter, every month,
> It bore its fair, well-ripened fruit;

There stood the tree alone, erect,
Its fruit than honey sweeter far,
That precious fruit so richly red
Did suffice for a man's nine meals ;
A year it added to man's life."
—Translated by Dr M'LAUCHLAN.

Page 26.

Ribes grossularia. The prickles of the gooseberry-bush were
used as charms for the cure of warts and the stye. A wedding-
ring laid over the wart, and pricked through the ring with a
gooseberry thorn, will remove the wart. Ten gooseberry thorns
are plucked to cure the stye—nine are pointed at the part
affected, and the tenth thrown over the left shoulder.

Page 31.

Meum athamanticum — *Muilceann.* The Inverness local
name for this plant, "*Bricin,*" is probably named after *St Bricin,*
who flourished about the year 637. He had a great establish-
ment at *Tuaim Dreeain.* His reputation as a saint and "*ollamh,*"
or doctor, extended far and wide; to him *Cennfaeladh,* the learned,
was carried to be cured after the battle of *Magh Rath.* He had
three schools for philosophy, classics, and law. It seems very
strange, however, that this local name should be confined to
Inverness, and be unknown in Ireland, where St Bricin was
residing.

Page 32.

Pastinaca sativa—(*Curran geal*) The white wild carrot,
parsnip. The natives of Harris make use of the seeds of the
wild white carrot, instead of hops, for brewing their beer, and
they say it answers the purpose sufficiently well, and gives the
drink a good relish besides.

"There is a large root growing amongst the rocks of this
island—the natives call it the ' *Curran petris,*' the rock-carrot
—of a whitish colour, and upwards of two feet in length, where
the ground is deep, and in shape and size like a large carrot."
—MARTIN.

Daucus carota—*Curran buidhe.* "The women present the
men (on St Michaelmas Day) with a pair of fine garters, of
divers colours, and they give them likewise a quantity of wild
carrots."—MARTIN.

Page 34.

Sambucus niger—(*Druman*) The elder. "The common people
[of the Highlands] keep as a great secret in curing wounds the

P

leaves of the elder, which they have gathered the first day of
April, for the purpose of disappointing the charms of witches.
They affix them to their doors and windows."—C. DE IRYNGIN,
at the Camp of Athole, June 30, 1651.

Misletoe and ivy were credited with similar powers. "The
inhabitants cut withies of misletoe and ivy, make circles of
them, keep them all the year, and pretend to cure hectic and
other troubles by them."—See Appendix to Pennant's 'Tour.'

"The misletoe," says Valancey, in his 'Grammar of the Irish
Language,' "was sacred to the Druids, because not only its
berries, but its leaves also, grew in clusters of three united to
one stock."

Page 38.

Carduus benedictus—*Fothannan beannuichte*, though applied
to "*Marianus*," is probably "*Centaurea benedictus*," and was
so called from the many medicinal virtues it was thought to
possess. It is a native of Spain and the Levant.

C. heterophyllus—Melancholy thistle. Was said to be the
badge of James I. of Scotland. A most appropriate badge;
but yet it had no connection with the unfortunate and melancholy history of the Stuarts, but was derived from the belief
that a decoction of this plant was a sovereign remedy for madness, which, in older times, was called "melancholy."

The plant generally selected to represent the Scotch heraldic
thistle is *Onopordon acanthium*, the cotton thistle, and, strange
to say, it does not grow wild in Scotland. Achaius, king of
Scotland (in the latter part of the eighth century), is said to
have been the first to have adopted the thistle for his device.
Favine says Achaius assumed the thistle in combination with
the rue: the thistle, because it will not endure handling; and
the rue, because it would drive away serpents by its smell,
and cure their poisonous bites. The thistle was not received
into the national arms before the fifteenth century.

Quercus robur—*Darach*. The age of the oak-tree was a
matter of much curiosity to the old Gaels :—

> " Tri aois coin, aois eich ;
> Tri aois eich, aois duine ;
> Tri aois duine, aois féidh ;
> Tri aois féidh, aois firein ;
> Tri aois firein, aois craoibh-dharaich."
>
> Thrice dog's age, age of horse.
> Thrice horse's age, age of man ;
> Thrice man's age, age of deer ;
> Thrice deer's age, age of eagle ;
> Thrice eagle's age, age of oak.

"The natives of Tiree preserve their yeast by an oaken wyth, which they twist and put into it, and for future use keep it in barley straw."—MARTIN.

<h3 style="text-align:center">Page 43.</h3>

Chrysanthemum leucanthemum — Ox eye daisy, called in Gaelic "*Breinean brothach.*" *Breinean* or *brainean* also means a king; Welsh, *brenhin.* The word is now obsolete in the Highlands. The plant was a remedy for the king's-evil.

<h3 style="text-align:center">Page 44.</h3>

Achillea millefolium—*Earr thalmhainn.* The yarrow, cut by moonlight by a young woman, with a black-handled knife, and certain mystic words, similar to the following, pronounced—

> "Good-morrow, good-morrow, fair yarrow,
> And thrice good-morrow to thee ;
> Come, tell me before to-morrow,
> Who my true love shall be."

The yarrow is brought home, put into the right stocking, and placed under the pillow, and the mystic dream is expected ; but if she opens her lips after she has pulled the yarrow, the charm is broken. Allusion is made to this superstition in a pretty song quoted in the 'Beauties of Highland Poetry,' p. 381, beginning—

> "Gu'n dh'eirich mi moch, air madainn an dé,
> 'S ghearr mi'n earr-thalmhainn, do bhri mo sgéil ;
> An dùil gu'm faicinn-sa rùin mo chléibh ;
> Ochòin ! gu'm facas, 's a cùl rium féin."
>
> I rose yesterday morning early,
> And cut the yarrow according to my skill,
> Expecting to see the beloved of my heart.
> Alas ! I saw him—but his back was towards me.

The superstitious customs described in Burns's "Halloween" were common among the Celtic races, and are more common on the western side of Scotland, from Galloway to Argyle, in consequence of that district having been occupied for centuries by the Dalriade Gaels.

<h3 style="text-align:center">Page 47.</h3>

Fraxinus excelsior—*Craobh uinnseann* (the ash-tree) was a most potent charm for cures of diseases of men and animals— *e.g.*, murrain in cattle, caused, it was supposed, by being stung in the mouth, or by being bitten by the larva of some moth. "Bore a hole in an ash-tree, and plug up the caterpillar in it, the leaves of that ash are a sure specific for that disease."

Martin adds, "the chief remedies were 'charms' for the cure of their diseases."

Page 51.

Verbena officinalis—*Trombhod.* Borlase, in his 'Antiquities of Cornwall,' speaking of the Druids, says: "They were excessively fond of the vervain; they used it in casting lots and foretelling events. It was gathered at the rising of the dogstar."

Page 68.

Corylus avellana—*Càlltuinn.* *Còl, càl,* in Welsh, signifies loss, also hazel-wood. The Welsh have a custom of presenting a forsaken lover with a stick of hazel, probably in allusion to the double meaning of the word.

Page 78.

Allium porrum—"*Bugha.*" The explanation given by Shaw that this was a name for leek seemed improbable, especially as it was a favourite comparison to the eye "when it is blue or dark." Turning to a passage describing Cormac Mac Airt, I found—

> "Cosmail ri *bugha* a shùili,"

which Professor O'Curry renders—

> "His eyes were like *slues,*"—

a far more appropriate comparison. Narcissus, *Lus a chròmchinn* (the bent head), suggests the beautiful lines of Herrick—

> "When a daffodill I see
> *Hanging its head* t'wards me,
> Guesse I may what I must be :
> First, I shall decline my head :
> Secondly, I shall be dead ;
> Lastly, safely burried."

Page 79.

A. ursinum — *Creamh.*

> "'Is leigheas air gach tinn
> Creamh 'us lm a' Mhaigh."
> Garlic and May butter
> Are remedies for every illness.

"Its medicinal virtues were well known; but like many other plants once valued and used by our ancestors, it is now quite superseded by pills and doses prepared by licensed practitioners."—Sheriff NICOLSON.

Page 81.

Potamogeton natans—*Duiliasg na h'aibhne.* The broad-leaved pondweed is used in connection with a curious superstition in some parts of Scotland, notably in the West Highlands. " It is gathered in small bundles in summer and autumn, where it is found to be plentiful, and kept until New Year's Day (old style); it is then put for a time into a tub or other dish of hot water, and the infusion is mixed with the first drink given to milch cows on New-Year's Day morning. This is supposed to keep the cows from witchcraft and the evil eye for the remainder of the year! It is also supposed to increase the yield of milk." —Rev. A. STEWART, Nether Lochaber.

Page 87.

Arundo phragmites—*Cruisgiornach* (*cruisigh*, in Irish, music, song). Reeds were said by the Greeks to have tended to sub-jugate nations by furnishing arrows for war, to soften their manners by means of music, and to lighten their understanding by supplying implements for writing. These modes of employment mark three different stages of civilisation. The great reed mace (*Typha latifolia*) *cuigeal nam bàn sithe*, is usually represented by painters in the hand of our Lord, as supposed to be the reed with which He was smitten by the Roman soldiers, and on which the sponge filled with vinegar was reached to Him.

Oats—*Coirc.* Martin mentions an ancient custom observed on the 2d of February. The mistress and servant of each family take a sheaf of oats and dress it in woman's apparel, put it in a large basket, with a wooden club by it, and this they call *Briid's* bed. They cry three times Briid is come, and welcome. This they do before going to bed, and when they rise in the morning they look at the ashes for the impress of Briid's club there; if seen, a prosperous year will follow.

Algæ—*Feamainn.* The inhabitants of the Isle of Lewis had an ancient custom of sacrificing to a sea-god called "Shony" at Hallowtide. The inhabitants round the island came to the church of St Mulvay, each person having provisions with him. One of their number was selected to wade into the sea up to the middle, and carrying a cup of ale in his hand, standing still in that position, crying out with a loud voice, "Shony, I give you this cup of ale, hoping you will be so kind as to send us plenty of sea-ware for enriching our ground the ensuing year." And he then threw the cup into the sea. This was performed in the night-time; they afterwards returned to spend the night in dancing and singing.

Shony (Sjoni), the Scandinavian Neptune. This offering was a relic of pagan worship introduced into the Western Isles by the Norwegians when they conquered and ruled over these islands centuries ago (*see* footnote, p. 40).

K'EOGH'S WORKS.—The Rev. John K'Eogh wrote a work on the plants of Ireland, 'Botanalogica Universalis Hibernia,' and another on the animals, 'Zoologica Medicinalis Hibernia,' about the year 1739, giving the Irish names as pronounced by the peasantry at that period. They are now rare works, and are of no value save for the names, for they contain no information except the supposed medicinal virtues of the plants and animals given in them.

All creatures, from the biggest mammal to the meanest worm, and all plants, were supposed to have some potent charm or virtue to cure disease. A large number of K'Eogh's prescriptions are compounds of the most disgusting ingredients. We can only now smile at the credulity that would lead any one to imagine that by merely looking at the yellow-hammer (*Emberiza citrinella*) "by any one who has the jaundice, the person is cured, but the bird will die." Or that "the eyes drawn entire out of the head of a hare taken in March, and dried with pepper, and worn by women, will facilitate childbirth."

He gives this singular cure for the jaundice: "A live moth, laid on the navel till it dies, is an excellent remedy! Nine grains of wheat *taken up by a flea*, are esteemed good to cure a chincough—that insect is banished and destroyed by elder leaves, flowers of pennyroyal, rue, mint, and fleabane, celandine, arsmart, mustard, brambles, lupin, and fern-root." For worms: "Take purslane seeds, coralina, and St John's-wort, of each an equal part; boil them in spring water. Or take of the powders of *hiera picra* (*Picris hieracioides*), of the seeds of the bitter apple, of each one dram, mixed with the oil of rue and savin, *spread on leather*, and apply it to the navel; this is an approved remedy." Epilepsy—"The flesh of the moor hen, with rosemary, lemons, lavender, and juniper berries, will cure it." And for children—"Take a whelp (*cullane*), a black sucking puppy (but a bitch whelp for a girl), strangle it, open it, and take out the gall, and give it to the child, and it will cure the falling-sickness." One more example will sufficiently illustrate the value of K'Eogh's books. "'Usnea capitis humani, or the moss growing on a skull that is exposed to the air, is a very good astringent, and stops bleeding if applied to the parts, *or even held in the hand.*"

Ollamh. This was the highest degree, in the ancient Gaelic system of learning, and before universities were established, included the study of law, medicine, poetry, classics, &c. A succession of such an order of *literati*, the Beatons, existed in Mull from time immemorial, until after the middle of last century. Their writings were all in Gaelic, to the amount of a large chestful. Dr Smith says that the remains of this treasure were bought as a literary curiosity for the library of the Duke of Chandos, and perished in the wreck of that nobleman's fortune. If this lost treasure could be recovered, we would have valuable material for a more complete collection of Gaelic names of plants, and information as to the uses to which they were applied, than we now possess.

MEDICINAL PLANTS.—The common belief that a plant grew not far from the locality where the disease prevailed, that would cure that disease, led to many experiments which ultimately resulted in finding out the undoubted virtues of many plants ; but wholesale methods were frequently adopted by gathering all the herbs, or as many as possible, in that particular district and making them into a bath.

At the battle of " Magh Tuireadh," we are informed " that the chief physician prepared a healing bath or fountain with the essences of the principal herbs and plants of Erinn, gathered chiefly in *Lus-Magh,* or the Plain of Herbs ; and on this bath they continued to pronounce incantations during the battle. Such of the men as happened to be wounded in the fight were immediately plunged into the bath, and they were instantly refreshed and made whole, so that they were able to return and fight against the enemy again and again."—Prof. O'CURRY.

INCANTATIONS WITH PLANTS.—Cures by incantations were most common. A large number of plants were thus employed. When John Roy Stewart sprained his ankle, when hiding after the battle of Culloden, he said :—

    " Ni mi'n ubhaidh rinn Peadar do Phàl,
    'S a lùighean air fàs leum bruaich,
    Seachd paidir n' ainm Sagairt a's Pàp
    Ga chuir ris na phlàsd mu'n cuairt."

I'll make the incantation that Peter made for Paul,
With the herbs that grew on the ground :
Seven paternosters in the name of priest and pope,
Applied like a plaster around.

" And if the dislocated joints did not at once jump into their
proper places during the recitation, the practitioner never failed
to augur favourably of the comfort to the patient. There were
similar incantations for all the ills that flesh is heir to : the
toothache could not withstand the potency of Highland magic ;
dysentery, gout, &c., had all their appropriate remedies in the
never-failing incantations."—M'KENZIE. See 'Beauties of Highland Poetry,' p. 268, where several of the " orations " repeated
as incantations arc given.

PLANTS AND FAIRY SUPERSTITIONS. — A large number of
plant - names in Gaelic have reference to fairy influence. At
births many ceremonies were used to baffle the fairy influence
over the child (see page 57), otherwise it would be carried off
to fairyland. The belief in fairies as well as most of these
superstitions, is traceable to the early ages of the British Druids,
on whose practices they are founded. The foxglove (*Meuran
sithe*), *odhran*, the cow-parsnip, and *copagach*, the docken, were
credited with great power in breaking the fairy spell; on the
other hand, some plants were supposed to facilitate the fairy
spell, and would cause the individual to be fairy " struck " or
"*buillite.*" The water-lily was supposed to possess this power,
hence its names, *Buillite*, and *Rabhagach*, meaning beware,
warning. Rushes found a place in fairy mythology : *Schœnus
nigricans (Seimhean)* furnished the shaft of the elf arrows, which
were tipped with white flint, and bathed in the dew that lies on
the hemlock.

NETTLES.—" They also used the roots of nettles and the roots
of reeds as cures for coughs." In some parts of Ireland there
is a custom on May eve and May day amongst the children,
especially the girls, of running amuck with branches of nettles,
stinging every one they meet. They had also a belief that steel
made hot and dipped in nettle-juice made it flexible. Camden
says " that the Romans cultivated nettles when in Britain in
order to rub their benumbed limbs with them, on account of the
intense cold they suffered when in Britain." A remedy worse
than the disease.

INDEX.

GAELIC NAMES.

Q

ENGLISH AND SCIENTIFIC.

CATALOGUE

OF

MESSRS BLACKWOOD & SONS'

PUBLICATIONS.

CATALOGUE

OF

MESSRS BLACKWOOD & SONS'

PUBLICATIONS.

ALISON. History of Europe. By Sir ARCHIBALD ALISON, Bart., D.C.L.

1. From the Commencement of the French Revolution to the Battle of Waterloo.
   LIBRARY EDITION, 14 vols., with Portraits. Demy 8vo, £10, 10s.
   ANOTHER EDITION, in 20 vols. crown 8vo, £6.
   PEOPLE'S EDITION, 13 vols. crown 8vo, £2, 11s.

2. Continuation to the Accession of Louis Napoleon.
   LIBRARY EDITION, 8 vols. 8vo, £6, 7s. 6d.
   PEOPLE'S EDITION, 8 vols. crown 8vo, 34s.

3. Epitome of Alison's History of Europe. Twenty-ninth Thousand, 7s. 6d.

4. Atlas to Alison's History of Europe. By A. Keith Johnston.
   LIBRARY EDITION, demy 4to, £3, 3s.
   PEOPLE'S EDITION. 31s. 6d.

—— Some Account of my Life and Writings: an Autobiography of the late Sir Archibald Alison, Bart., D.C.L. Edited by his Daughter-in-law. 2 vols. 8vo, with Portrait engraved on Steel. [*In the Press.*

—— Life of John Duke of Marlborough. With some Account of his Contemporaries, and of the War of the Succession. Third Edition, 2 vols. 8vo. Portraits and Maps, 30s.

—— Essays : Historical, Political, and Miscellaneous. 3 vols. demy 8vo, 45s.

—— Lives of Lord Castlereagh and Sir Charles Stewart, Second and Third Marquesses of Londonderry. From the Original Papers of the Family. 3 vols. 8vo, £2, 2s.

—— Principles of the Criminal Law of Scotland. 8vo, 18s.

—— Practice of the Criminal Law of Scotland. 8vo, cloth boards, 18s.

—— The Principles of Population, and their Connection with Human Happiness. 2 vols. 8vo, 30s.

ALISON. On the Management of the Poor in Scotland, and its Effects on the Health of the Great Towns. By WILLIAM PULTENEY ALISON, M.D. Crown 8vo, 5s. 6d.

ADAMS. Great Campaigns. A Succinct Account of the Principal Military Operations which have taken place in Europe from 1796 to 1870. By Major C. ADAMS, Professor of Military History at the Staff College. Edited by Captain C. COOPER KING, R.M. Artillery, Instructor of Tactics, Royal Military College. 8vo, with Maps. 16s.

AIRD. Poetical Works of Thomas Aird. Fifth Edition, with
Memoir of the Author by the Rev. JARDINE WALLACE, and Portrait.
Crown 8vo, 7s. 6d.

——— The Old Bachelor in the Old Scottish Village. Fcap. 8vo, 4s.

ALLARDYCE. The City of Sunshine. By ALEXANDER ALLAR-
DYCE. Three vols. post 8vo, £1, 5s. 6d.

——— Memoir of the Honourable George Keith Elphinstone,
K.B., Viscount Keith of Stonehaven Marischal, Admiral of the Red. One vol.
8vo, with Portrait, Illustrations, and Maps.   [In the Press.

ANCIENT CLASSICS FOR ENGLISH READERS. Edited by
Rev. W. LUCAS COLLINS, M.A. Complete in 28 vols., cloth, 2s. 6d. each ; or in
14 vols., tastefully bound, with calf or vellum back, £3, 10s.

Contents of the Series.

HOMER: THE ILIAD. By the Editor.	PLAUTUS AND TERENCE. By the Editor.
HOMER: THE ODYSSEY. By the Editor.	THE COMMENTARIES OF CÆSAR. By An-
HERODOTUS. By George C. Swayne,	thony Trollope.
M.A.	TACITUS. By W. B. Donne.
XENOPHON. By Sir Alexander Grant,	CICERO. By the Editor.
Bart., LL.D.	PLINY'S LETTERS. By the Rev. Alfred
EURIPIDES. By W. B. Donne.	Church, M.A., and the Rev. W. J. Brod-
ARISTOPHANES. By the Editor.	ribb, M.A.
PLATO. By Clifton W. Collins, M.A.	LIVY. By the Editor.
LUCIAN. By the Editor.	OVID. By the Rev. A. Church, M.A.
ÆSCHYLUS. By the Right Rev. the Bishop	CATULLUS, TIBULLUS, AND PROPERTIUS.
of Colombo.	By the Rev. Jas. Davies, M.A.
SOPHOCLES. By Clifton W. Collins, M.A.	DEMOSTHENES. By the Rev. W. J. Brod-
HESIOD AND THEOGNIS. By the Rev. J.	ribb, M.A.
Davies, M.A.	ARISTOTLE. By Sir Alexander Grant,
GREEK ANTHOLOGY. By Lord Neaves.	Bart., LL.D.
VIRGIL. By the Editor.	THUCYDIDES. By the Editor.
HORACE. By Sir Theodore Martin, K.C.B.	LUCRETIUS. By W. H. Mallock, M.A.
JUVENAL. By Edward Walford, M.A.	PINDAR. By the Rev. F. D. Morice, M.A.

AYLWARD. The Transvaal of To-day : War, Witchcraft,
Sports, and Spoils in South Africa. By ALFRED AYLWARD, Commandant,
Transvaal Republic ; Captain (late) Lydenberg Volunteer Corps. Second
Edition. Crown 8vo, with a Map, 6s.

AYTOUN. Lays of the Scottish Cavaliers, and other Poems. By
W. EDMONDSTOUNE AYTOUN, D.C.L., Professor of Rhetoric and Belles-Lettres
in the University of Edinburgh. Twenty-eighth Edition. Fcap. 8vo, 7s. 6d.

——— An Illustrated Edition of the Lays of the Scottish Cavaliers.
From designs by Sir NOEL PATON. Small 4to, 21s., in gilt cloth.

——— Bothwell : a Poem. Third Edition. Fcap., 7s. 6d.

——— Firmilian ; or, The Student of Badajoz. A Spasmodic
Tragedy. Fcap., 5s.

——— Poems and Ballads of Goethe. Translated by Professor
AYTOUN and Sir THEODORE MARTIN, K.C.B. Third Edition. Fcap., 6s.

——— Bon Gaultier's Book of Ballads. By the SAME. Thirteenth
Edition. With Illustrations by Doyle, Leech, and Crowquill. Post 8vo, gilt
edges, 8s. 6d.

——— The Ballads of Scotland. Edited by Professor AYTOUN.
Fourth Edition. 2 vols. fcap. 8vo, 12s.

——— Memoir of William E. Aytoun, D.C.L. By Sir THEODORE
MARTIN, K.C.B. With Portrait. Post 8vo, 12s.

BAGOT. The Art of Poetry of Horace. Free and Explanatory
Translations in Prose and Verse. By the Very Rev. DANIEL BAGOT, D.D.
Third Edition, Revised, printed on *papier vergé*, square 8vo, 5s.

BAIRD LECTURES. The Mysteries of Christianity. By T. J. CRAWFORD, D.D., F.R.S.E., Professor of Divinity in the University of Edinburgh, &c. Being the Baird Lecture for 1874. Crown 8vo, 7s. 6d.

—— Endowed Territorial Work : Its Supreme Importance to the Church and Country. By WILLIAM SMITH, D.D., Minister of North Leith. Being the Baird Lecture for 1875. Crown 8vo, 6s.

—— Theism. By ROBERT FLINT, D.D., LL.D., Professor of Divinity in the University of Edinburgh. Being the Baird Lecture for 1876. Third Edition. Crown 8vo, 7s. 6d.

—— Anti-Theistic Theories. By the SAME. Being the Baird Lecture for 1877. Second Edition. Crown 8vo, 10s. 6d.

BATTLE OF DORKING. Reminiscences of a Volunteer. From 'Blackwood's Magazine.' Second Hundredth Thousand. 6d.

BY THE SAME AUTHOR.

The Dilemma. Cheap Edition. Crown 8vo, 6s.

BEDFORD. The Regulations of the Old Hospital of the Knights of St John at Valetta. From a Copy Printed at Rome, and preserved in the Archives of Malta; with a Translation, Introduction, and Notes Explanatory of the Hospital Work of the Order. By the Rev. W. K. R. BEDFORD, one of the Chaplains of the Order of St John in England. Royal 8vo, with Frontispiece, Plans, &c., 7s. 6d.

BESANT. Readings from Rabelais. By WALTER BESANT, M.A. In one volume, post 8vo. [In the press.

BLACKIE. Lays and Legends of Ancient Greece. By JOHN STUART BLACKIE, Professor of Greek in the University of Edinburgh. Second Edition. Fcap. 8vo. 5s.

BLACKWOOD'S MAGAZINE, from Commencement in 1817 to June 1880. Nos. 1 to 776, forming 127 Volumes.

—— Index to Blackwood's Magazine. Vols. 1 to 50. 8vo, 15s.

—— Tales from Blackwood. Forming Twelve Volumes of Interesting and Amusing Railway Reading. Price One Shilling each in Paper Cover. Sold separately at all Railway Bookstalls.

They may also be had bound in cloth, 18s., and in half calf, richly gilt, 30s. or 12 volumes in 6, Roxburghe, 21s., and half red morocco, 28s.

—— Tales from Blackwood. New Series. Complete in Twenty-four Shilling Parts. Handsomely bound in 12 vols., cloth, 30s. In leather back, Roxburghe style, 37s. 6d. In half calf, gilt, 52s. 6d. In half morocco, 55s.

—— Standard Novels. Uniform in size and legibly Printed. Each Novel complete in one volume.

Florin Series, Illustrated Boards.

TOM CRINGLE'S LOG. By Michael Scott.	PEN OWEN. By Dean Hook.
THE CRUISE OF THE MIDGE. By the Same.	ADAM BLAIR. By J. G. Lockhart.
CYRIL THORNTON. By Captain Hamilton.	LADY LEE'S WIDOWHOOD. By General
ANNALS OF THE PARISH. By John Galt.	Sir E. B. Hamley.
THE PROVOST, &c. By John Galt.	SALEM CHAPEL. By Mrs Oliphant.
SIR ANDREW WYLIE. By John Galt.	THE PERPETUAL CURATE. By Mrs Oli-
THE ENTAIL. By John Galt.	phant.
MISS MOLLY. By Beatrice May Butt.	MISS MARJORIBANKS. By Mrs Oliphant.
REGINALD DALTON. By J. G. Lockhart.	JOHN : A Love Story. By Mrs Oliphant.

Or in Cloth Boards, 2s. 6d.

Shilling Series, Illustrated Cover.

THE RECTOR, and THE DOCTOR'S FAMILY. By Mrs Oliphant.	SIR FRIZZLE PUMPKIN, NIGHTS AT MESS, &c.
THE LIFE OF MANSIE WAUCH. By D. M. Moir.	THE SUBALTERN.
PENINSULAR SCENES AND SKETCHES. By F. Hardman.	LIFE IN THE FAR WEST. By G. F. Ruxton.
	VALERIUS : A Roman Story. By J. G. Lockhart.

Or in Cloth Boards, 1s. 6d.

BLACKMORE.   The Maid of Sker.   By R. D. BLACKMORE, Author
of 'Lorna Doone,' &c.   Eighth Edition.   Crown 8vo, 7s. 6d.

BOSCOBEL TRACTS.   Relating to the Escape of Charles the
Second after the Battle of Worcester, and his subsequent Adventures.   Edited
by J. HUGHES, Esq., A.M.   A New Edition, with additional Notes and Illustrations, including Communications from the Rev. R. H. BARHAM, Author of
the 'Ingoldsby Legends.'   8vo, with Engravings, 16s.

BRACKENBURY.   A Narrative of the Ashanti War.   Prepared
from the official documents, by permission of Major-General Sir Garnet Wolseley, K.C.B., K.C.M.G.   By Major H. BRACKENBURY, R.A., Assistant Military
Secretary to Sir Garnet Wolseley.   With Maps from the latest Surveys made by
the Staff of the Expedition.   2 vols. 8vo, 25s.

BROADLEY.   Tunis, Past and Present.   By A. M. BROADLEY.
With numerous Illustrations and Maps.   2 vols. post 8vo.   [In the Press.

BROOKE, Life of Sir James, Rajah of Sarāwak.   From his Personal
Papers and Correspondence.   By SPENSER ST JOHN, H.M.'s Minister-Resident
and Consul-General Peruvian Republic; formerly Secretary to the Rajah.
With Portrait and a Map.   Post 8vo, 12s. 6d.

BROUGHAM.   Memoirs of the Life and Times of Henry Lord
Brougham.   Written by HIMSELF.   3 vols. 8vo, £2, 8s.   The Volumes are sold
separately, price 16s. each.

BROWN.   The Forester: A Practical Treatise on the Planting,
Rearing, and General Management of Forest-trees.   By JAMES BROWN, Wood-
Surveyor and Nurseryman.   Fifth Edition, revised and enlarged.   Royal
8vo, with Engravings.   [Nearly ready.

BROWN.   The Ethics of George Eliot's Works.   By JOHN CROMBIE
BROWN.   Third Edition.   Crown 8vo, 2s. 6d.

BROWN.   A Manual of Botany, Anatomical and Physiological.
For the Use of Students.   By ROBERT BROWN, M.A., Ph.D., F.L.S., F.R.G.S.
Crown 8vo, with numerous Illustrations, 12s. 6d.

BUCHAN.   Introductory Text-Book of Meteorology.   By ALEX-
ANDER BUCHAN, M.A., F.R.S.E., Secretary of the Scottish Meteorological
Society, &c.   Crown 8vo, with 8 Coloured Charts and other Engravings,
pp. 218.   4s. 6d.

BURBIDGE.   Domestic Floriculture, Window Gardening, and
Floral Decorations.   Being practical directions for the Propagation, Culture,
and Arrangement of Plants and Flowers as Domestic Ornaments.   By F. W.
BURBIDGE.   Second Edition.   Crown 8vo, with numerous Illustrations, 7s. 6d.

—— Cultivated Plants: Their Propagation and Improvement.
Including Natural and Artificial Hybridisation, Raising from Seed, Cuttings,
and Layers, Grafting and Budding, as applied to the Families and Genera in
Cultivation.   Crown 8vo, with numerous Illustrations, 12s. 6d.

BURN.   Handbook of the Mechanical Arts Concerned in the Con-
struction and Arrangement of Dwelling-Houses and other Buildings; with
Practical Hints on Road-making and the Enclosing of Land.   By ROBERT SCOTT
BURN, Engineer.   Second Edition.   Crown 8vo, 6s. 6d.

BURTON.   The History of Scotland: From Agricola's Invasion to
the Extinction of the last Jacobite Insurrection.   By JOHN HILL BURTON,
D.C.L., Historiographer-Royal for Scotland.   New and Enlarged Edition,
8 vols., and Index.   Crown 8vo, £3, 3s.

—— History of the British Empire during the Reign of Queen
Anne.   In 3 vols. 8vo.   36s.

—— The Cairngorm Mountains.   Crown 8vo, 3s. 6d.

—— The Scot Abroad.   Second Edition.   Complete in One volume.   Crown 8vo, 10s. 6d.

—— The Book-Hunter.   A New and Choice Edition.   With a
Memoir of the Author, a Portrait etched by Mr Hole, A.R.S.A., and other
Illustrations.   In small 4to, on hand-made paper.   [In the Press.

BUTE. The Roman Breviary: Reformed by Order of the Holy
Œcumenical Council of Trent; Published by Order of Pope St Pius V.; and
Revised by Clement VIII. and Urban VIII.; together with the Offices since
granted. Translated out of Latin into English by JOHN, Marquess of Bute,
K.T. In 2 vols. crown 8vo, cloth boards, edges uncut. £2, 2s.

—— The Altus of St Columba. With a Prose Paraphrase and
Notes. In paper cover, 2s. 6d.

BUTT. Miss Molly. By BEATRICE MAY BUTT. Cheap Edition, 2s.

—— Delicia. By the Author of 'Miss Molly.' Fourth Edition. Crown 8vo, 7s. 6d.

CAIRD. Sermons. By JOHN CAIRD, D.D., Principal of the University of Glasgow. Fourteenth Thousand. Fcap. 8vo, 5s.

—— Religion in Common Life. A Sermon preached in Crathie
Church, October 14, 1855, before Her Majesty the Queen and Prince Albert.
Published by Her Majesty's Command Cheap Edition, 3d.

CAMPBELL, Life of Colin, Lord Clyde. *See* General SHADWELL,
at page 20.

CAMPBELL. Sermons Preached before the Queen at Balmoral.
By the Rev. A. A. CAMPBELL, Minister of Crathie. Published by Command
of Her Majesty. Crown 8vo, 4s. 6d.

CARLYLE. Autobiography of the Rev. Dr Alexander Carlyle,
Minister of Inveresk. Containing Memorials of the Men and Events of his
Time. Edited by JOHN HILL BURTON. 8vo. Third Edition, with Portrait, 14s.

CARRICK. Koumiss; or, Fermented Mare's Milk: and its Uses
in the Treatment and Cure of Pulmonary Consumption, and other Wasting
Diseases. With an Appendix on the best Methods of Fermenting Cow's Milk.
By GEORGE L. CARRICK, M.D., L.R.C.S.E. and L.R.C.P.E., Physician to the
British Embassy, St Petersburg, &c. Crown 8vo, 10s. 6d.

CAUVIN. A Treasury of the English and German Languages.
Compiled from the best Authors and Lexicographers in both Languages.
Adapted to the Use of Schools, Students, Travellers, and Men of Business;
and forming a Companion to all German-English Dictionaries. By JOSEPH
CAUVIN, LL.D. & Ph.D., of the University of Göttingen, &c. Crown 8vo,
7s. 6d.

CAVE-BROWN. Lambeth Palace and its Associations. By J.
CAVE-BROWN, M.A., Vicar of Detling, Kent, and for many years Curate of Lam-
beth Parish Church. With an Introduction by the Archbishop of Canterbury.
In One volume, with Illustrations. [*In the Press.*

CHARTERIS. Canonicity; or, Early Testimonies to the Existence
and Use of the Books of the New Testament. Based on Kirchhoffer's 'Quel-
lensammlung.' Edited by A. H. CHARTERIS, D.D., Professor of Biblical
Criticism in the University of Edinburgh. 8vo, 18s.

—— Life of the Rev. James Robertson, D.D., F.R.S.E., Professor of Divinity and Ecclesiastical History in the University of Edinburgh.
By Professor CHARTERIS. With Portrait. 8vo. 10s. 6d.

CHEVELEY NOVELS, THE.
I. A MODERN MINISTER. 2 vols. bound in cloth, with Twenty-six Illustrations.
17s.
II. SAUL WEIR. 2 vols. bound in cloth. With Twelve Illustrations by F. Bar-
nard. 16s.

CHIROL. 'Twixt Greek and Turk. By M. VALENTINE CHIROL.
Post 8vo. With Frontispiece and Map, 10s. 6d.

CHURCH SERVICE SOCIETY. A Book of Common Order:
Being Forms of Worship issued by the Church Service Society. Fourth Edition, 5s.

COLQUHOUN. The Moor and the Loch. Containing Minute
Instructions in all Highland Sports, with Wanderings over Crag and Corrie,
Flood and Fell. By John Colquhoun. Fifth Edition, greatly enlarged.
With Illustrations. 2 vols. post 8vo, 26s.

COTTERILL. The Genesis of the Church. By the Right. Rev.
Henry Cotterill, D.D., Bishop of Edinburgh. Demy 8vo, 16s.

CRANSTOUN. The Elegies of Albius Tibullus. Translated into
English Verse, with Life of the Poet, and Illustrative Notes. By James Cranstoun, LL.D., Author of a Translation of 'Catullus.' Crown 8vo, 6s. 6d.

—— The Elegies of Sextus Propertius. Translated into English
Verse, with Life of the Poet, and Illustrative Notes. Crown 8vo, 7s. 6d.

CRAWFORD. The Doctrine of Holy Scripture respecting the
Atonement. By the late Thomas J. Crawford, D.D., Professor of Divinity in
the University of Edinburgh. Third Edition. 8vo, 12s.

—— The Fatherhood of God, Considered in its General and
Special Aspects, and particularly in relation to the Atonement, with a
Review of Recent Speculations on the Subject. Third Edition, Revised and
Enlarged. 8vo, 9s.

—— The Preaching of the Cross, and other Sermons. 8vo,
7s. 6d.

—— The Mysteries of Christianity; being the Baird Lecture for
1874. Crown 8vo, 7s. 6d.

CROSSE. Round about the Carpathians. By Andrew F. Crosse,
F.C.S. 8vo, with Map of the Author's route, price 12s. 6d.

DESCARTES. The Method, Meditations, and Principles of Philo-
sophy of Descartes. Translated from the Original French and Latin. With a
New Introductory Essay, Historical and Critical, on the Cartesian Philosophy.
By John Veitch, LL.D., Professor of Logic and Rhetoric in the University of
Glasgow. A New Edition, during the Eighth. Price 6s. 6d.

DICKSON. Japan; being a Sketch of the History, Government,
and Officers of the Empire. By Walter Dickson. 8vo, 15s.

DU CANE. The Odyssey of Homer, Books I.-XII. Translated into
English Verse. By Sir Charles Du Cane, K.C.M.G. 8vo, 10s. 6d.

DUDGEON. History of the Edinburgh or Queen's Regiment
Light Infantry Militia, now Third Battalion, The Royal Scots; with an
Account of the Origin and Progress of the Militia, and a Brief Sketch of the
old Royal Scots. By Major R. C. Dudgeon, Adjutant 3d Battalion The Royal
Scots. Post 8vo, with Illustrations, 10s. 6d.

EAGLES. Essays. By the Rev. John Eagles, A.M. Oxon. Originally published in 'Blackwood's Magazine.' Post 8vo, 10s. 6d.

—— The Sketcher. Originally published in 'Blackwood's
Magazine.' Post 8vo, 10s. 6d.

ELIOT. Impressions of Theophrastus Such. By George Eliot.
New and cheaper Edition. Crown 8vo, 5s.

—— Adam Bede. Illustrated Edition. 3s. 6d., cloth.

—— The Mill on the Floss. Illustrated Edition. 3s. 6d., cloth.

—— Scenes of Clerical Life. Illustrated Edition. 3s., cloth.

—— Silas Marner: The Weaver of Raveloe. Illustrated Edition. 2s. 6d., cloth.

—— Felix Holt, the Radical. Illustrated Edition. 3s. 6d., cloth.

—— Romola. With Vignette. 3s. 6d., cloth.

—— Middlemarch. Crown 8vo, 7s. 6d.

—— Daniel Deronda. Crown 8vo, 7s. 6d.

ELIOT. Works of George Eliot (Cabinet Edition). Complete and
Uniform Edition, handsomely printed in a new type, 20 volumes, crown 8vo,
price £5. The Volumes are also sold separately, price 5s. each, viz.:—
Romola. 2 vols.—Silas Marner, The Lifted Veil, Brother Jacob. 1 vol.—
Adam Bede. 2 vols.—Scenes of Clerical Life. 2 vols.—The Mill on
the Floss. 2 vols.—Felix Holt. 2 vols.—Middlemarch. 3 vols.—
Daniel Deronda. 3 vols.—The Spanish Gypsy. 1 vol.—Jubal, and
other Poems, Old and New. 1 vol.—Theophrastus Such. 1 vol.

—— The Spanish Gypsy. Crown 8vo, 5s.

—— The Legend of Jubal, and other Poems, Old and New.
New Edition. Fcap. 8vo, 5s., cloth.

—— Wise, Witty, and Tender Sayings, in Prose and Verse.
Selected from the Works of GEORGE ELIOT. Fifth Edition. Fcap. 8vo, 6s.

—— The George Eliot Birthday Book. Printed on fine paper,
with red border, and handsomely bound in cloth, gilt. Fcap. 8vo, cloth, 3s. 6d.
And in French morocco or Russia, 5s.

ESSAYS ON SOCIAL SUBJECTS. Originally published in
the 'Saturday Review.' A New Edition. First and Second Series. 2 vols.
crown 8vo, 6s. each.

EWALD. The Crown and its Advisers; or, Queen, Ministers,
Lords, and Commons. By ALEXANDER CHARLES EWALD, F.S.A. Crown 8vo,
5s.

THE FAITHS OF THE WORLD. A Concise History of the
Great Religious Systems of the World. By various Authors. Being the St
Giles' Lectures—Second Series. Complete in One Volume, Crown 8vo, 5s.
Sold separately, price 4d.

FERGUSSON. The Honourable Henry Erskine, Lord Advocate
for Scotland. With Notices of certain of his Kinsfolks and of his Time
Compiled from Family Papers, and other sources of Information. By LIEU-
TENANT-COLONEL ALEX. FERGUSSON, late of the Staff of her Majesty's Indian
Army. In One volume, large 8vo. With Portraits and other Illustra-
tions. [Shortly.

FERRIER. Philosophical Works of the late James F. Ferrier,
B.A. Oxon., Professor of Moral Philosophy and Political Economy, St Andrews.
New Edition. Edited by Sir ALEX. GRANT, Bart., D.C.L., and Professor
LUSHINGTON. 3 vols. crown 8vo, 34s. 6d.

—— Institutes of Metaphysic. Third Edition. 10s. 6d.

—— Lectures on the Early Greek Philosophy. Third Edition.
10s. 6d.

—— Philosophical Remains, including the Lectures on Early
Greek Philosophy. 2 vols., 24s.

FERRIER. George Eliot and Judaism. An Attempt to appreciate
'Daniel Deronda.' By Professor DAVID KAUFMANN, of the Jewish Theological
Seminary, Buda-Pesth. Translated from the German by J. W. FERRIER.
Second Edition. Crown 8vo, 2s. 6d.

FLINT. The Philosophy of History in Europe. Vol. I., containing the History of that Philosophy in France and Germany. By ROBERT FLINT,
D.D., LL.D., Professor of Divinity, University of Edinburgh. 8vo, 15s.

—— Theism. Being the Baird Lecture for 1876. Third Edition.
Crown 8vo, 7s. 6d.

—— Anti-Theistic Theories. Being the Baird Lecture for 1877.
Second Edition. Crown 8vo, 10s. 6d.

FORBES. The Campaign of Garibaldi in the Two Sicilies: A Personal Narrative. By CHARLES STUART FORBES, Commander, R.N. Post 8vo,
with Portraits, 12s.

FOREIGN CLASSICS FOR ENGLISH READERS. Edited by Mrs OLIPHANT. Price 2s. 6d.

Now published:—

DANTE. By the Editor.
VOLTAIRE. By General Sir E. B. Hamley, K.C.M.G.
PASCAL. By Principal Tulloch.
PETRARCH. By Henry Reeve, C.B.
GOETHE. By A. Hayward, Q.C.
MOLIÈRE. By the Editor and F. Tarver, M.A.
MONTAIGNE. By Rev. W. L. Collins, M.A.
RABELAIS. By Walter Besant, M.A.
CALDERON. By E. J. Hasell.

SAINT SIMON. By Clifton W. Collins, M.A.
CERVANTES. By the Editor.
CORNEILLE AND RACINE. By Henry M. Trollope.
MADAME DE SÉVIGNÉ. By Miss Thackeray.
LA FONTAINE, AND OTHER FRENCH FABULISTS. By Rev. W. L. Collins, M.A.
SCHILLER. By James Sime, M.A., Author of 'Lessing: his Life and Writings.'

In preparation:—ROUSSEAU. By Henry Graham.—TASSO. By E. J. Hasell.

FRASER. Handy Book of Ornamental Conifers, and of Rhododendrons and other American Flowering Shrubs, suitable for the Climate and Soils of Britain. With descriptions of the best kinds, and containing Useful Hints for their successful Cultivation. By HUGH FRASER, Fellow of the Botanical Society of Edinburgh. Crown 8vo, 6s.

GALT. Annals of the Parish. By JOHN GALT. Fcap. 8vo, 2s.

——— The Provost. Fcap. 8vo, 2s.

——— Sir Andrew Wylie. Fcap. 8vo, 2s.

——— The Entail ; or, The Laird of Grippy. Fcap. 8vo, 2s.

GARDENER, THE: A Magazine of Horticulture and Floriculture. Edited by DAVID THOMSON, Author of 'The Handy Book of the Flower-Garden,' &c.; Assisted by a Staff of the best practical Writers. Published Monthly, 6d.

GENERAL ASSEMBLY OF THE CHURCH OF SCOTLAND.
——— Family Prayers. Authorised by the General Assembly of the Church of Scotland. A New Edition, crown 8vo, in large type, 4s. 6d. Another Edition, crown 8vo, 2s.

——— Prayers for Social and Family Worship. For the Use of Soldiers, Sailors, Colonists, and Sojourners in India, and other Persons, at home and abroad, who are deprived of the ordinary services of a Christian Ministry. Cheap Edition, 1s. 6d.

——— The Scottish Hymnal. Hymns for Public Worship. Published for Use in Churches by Authority of the General Assembly. Various sizes—viz.: 1. Large type, for pulpit use, cloth, 3s. 6d. 2. Longprimer type, cloth, red edges, 1s. 6d.; French morocco, 2s. 6d.; calf, 6s. 3. Bourgeois type, cloth, red edges, 1s.; French morocco, 2s. 4. Minion type, limp cloth, 6d.; French morocco, 1s. 6d. 5. School Edition, in paper cover, 2d. 6. Children's Hymnal, paper cover, 1d. No. 2, bound with the Psalms and Paraphrases, cloth, 3s.; French morocco, 4s. 6d.; calf, 7s. 6d. No. 3, bound with the Psalms and Paraphrases, cloth, 2s. ; French morocco, 3s.

——— The Scottish Hymnal, with Music. Selected by the Committees on Hymns and on Psalmody. The harmonies arranged by W. H. Monk. Cloth, 1s. 6d.; French morocco, 3s. 6d. The same in the Tonic Sol-fa Notation, 1s. 6d. and 3s. 6d.

——— The Scottish Hymnal, with Fixed Tune for each Hymn. Longprimer type, 3s. 6d.

GERARD. Reata: What's in a Name? By E. D. GERARD. New Edition. In one volume, crown 8vo, 6s.

——— Beggar my Neighbour. A Novel. New Edition, complete in one volume, crown 8vo, 6s.

GLEIG. The Subaltern. By G. R. GLEIG, M.A., late Chaplain-General of Her Majesty's Forces. Originally published in 'Blackwood's Magazine.' Library Edition. Revised and Corrected, with a New Preface. Crown 8vo, 7s. 6d.

GOETHE'S FAUST. Translated into English Verse by Sir THEODORE MARTIN, K.C.B. Second Edition, post 8vo, 6s. Cheap Edition, fcap., 3s. 6d.

—— Poems and Ballads of Goethe. Translated by Professor AYTOUN and Sir THEODORE MARTIN, K.C.B. Third Edition, fcap. 8vo, 6s.

GORDON CUMMING. At Home in Fiji. By C. F. GORDON CUMMING, Author of 'From the Hebrides to the Himalayas.' Fourth Edition, complete in one volume post 8vo. With Illustrations and Map. 7s. 6d.

—— A Lady's Cruise in a French Man-of-War. New and Cheaper Edition. In one volume, 8vo. With Illustrations and Map. 12s. 6d.

GRAHAM. Annals and Correspondence of the Viscount and First and Second Earls of Stair. By JOHN MURRAY GRAHAM. 2 vols. demy 8vo, with Portraits and other Illustrations. £1, 8s.

—— Memoir of Lord Lynedoch. Second Edition, crown 8vo, 5s.

GRANT. Bush-Life in Queensland. By A. C. GRANT. New Edition In One volume crown 8vo, 6s.

GRANT. Incidents in the Sepoy War of 1857-58. Compiled from the Private Journals of the late General Sir HOPE GRANT, G.C.B. ; together with some Explanatory Chapters by Captain HENRY KNOLLYS, R.A. Crown 8vo, with Map and Plans, 12s.

GRANT. Memorials of the Castle of Edinburgh. By JAMES GRANT. A New Edition. Crown 8vo, with 12 Engravings, 2s.

HAMERTON. Wenderholme : A Story of Lancashire and Yorkshire Life. By PHILIP GILBERT HAMERTON, Author of 'A Painter's Camp.' A New Edition. Crown 8vo, 6s.

HAMILTON. Lectures on Metaphysics. By Sir WILLIAM HAMILTON, Bart., Professor of Logic and Metaphysics in the University of Edinburgh. Edited by the Rev. H. L. MANSEL, B.D., LL.D., Dean of St Paul's ; and JOHN VEITCH, M.A., Professor of Logic and Rhetoric, Glasgow. Sixth Edition. 2 vols. 8vo, 24s.

—— Lectures on Logic. Edited by the SAME. Third Edition. 2 vols. 24s.

—— Discussions on Philosophy and Literature, Education and University Reform. Third Edition, 8vo, 21s.

—— Memoir of Sir William Hamilton, Bart., Professor of Logic and Metaphysics in the University of Edinburgh. By Professor VEITCH of the University of Glasgow. 8vo, with Portrait, 18s.

HAMILTON. Annals of the Peninsular Campaigns. By Captain THOMAS HAMILTON. Edited by F. Hardman. 8vo, 16s. Atlas of Maps to illustrate the Campaigns, 12s.

HAMLEY. The Operations of War Explained and Illustrated. By General Sir EDWARD BRUCE HAMLEY, K.C.M.G. Fourth Edition, revised throughout. 4to, with numerous Illustrations, 30s.

—— Thomas Carlyle : An Essay. Second Edition. Crown 8vo. 2s. 6d.

—— The Story of the Campaign of Sebastopol. Written in the Camp. With Illustrations drawn in Camp by the Author. 8vo, 21s.

—— On Outposts. Second Edition. 8vo, 2s.

—— Wellington's Career ; A Military and Political Summary. Crown 8vo, 2s.

—— Lady Lee's Widowhood. Crown 8vo, 2s. 6d.

HAMLEY. Our Poor Relations. A Philozoic Essay. With Illustrations, chiefly by Ernest Griset. Crown 8vo, cloth gilt, 3s. 6d.

HAMLEY. Guilty, or Not Guilty? A Tale. By Major-General W. G. HAMLEY, late of the Royal Engineers. New Edition. Crown 8vo, 3s. 6d.

——— The House of Lys : One Book of its History. A Tale. Second Edition. 2 vols. crown 8vo. 17s.

——— Traseaden Hall. "When George the Third was King." 3 vols., post 8vo, 25s. 6d.

HANDY HORSE-BOOK ; or, Practical Instructions in Riding, Driving, and the General Care and Management of Horses. By 'MAGENTA.' Ninth Edition, with 6 Engravings, 4s. 6d.

BY THE SAME.

Our Domesticated Dogs : their Treatment in reference to Food, Diseases, Habits, Punishment, Accomplishments. Crown 8vo, 2s. 6d.

HARBORD. Definitions and Diagrams in Astronomy and Navigation. By the Rev. J. B. HARBORD, M.A., Assistant Director of Education, Admiralty. 1s.

——— Short Sermons for Hospitals and Sick Seamen. Fcap. 8vo, cloth, 4s. 6d.

HARDMAN. Scenes and Adventures in Central America. Edited by FREDERICK HARDMAN. Crown 8vo, 6s.

HAWKEY. The Shakespeare Tapestry. Woven in Verse. By C. HAWKEY. Fcap. 8vo. 6s.

HAY. The Works of the Right Rev. Dr George Hay, Bishop of Edinburgh. Edited under the Supervision of the Right Rev. Bishop STRAIN. With Memoir and Portrait of the Author. 5 vols. crown 8vo, bound in extra cloth, £1, 1s. Or, sold separately—viz. :

——— The Sincere Christian Instructed in the Faith of Christ from the Written Word. 2 vols., 8s.

——— The Devout Christian Instructed in the Law of Christ from the Written Word. 2 vols., 8s.

——— The Pious Christian Instructed in the Nature and Practice of the Principal Exercises of Piety. 1 vol., 4s.

HEATLEY. The Horse-Owner's Safeguard. A Handy Medical Guide for every Man who owns a Horse. By G. S. HEATLEY, V.S In one volume, crown 8vo. [In the Press.

HEMANS. The Poetical Works of Mrs Hemans. Copyright Editions.
One Volume, royal 8vo, 5s.
The Same, with Illustrations engraved on Steel, bound in cloth, gilt edges, 7s. 6d.
Six Volumes in Three, fcap., 12s. 6d.
SELECT POEMS OF MRS HEMANS. Fcap., cloth, gilt edges, 3s.

——— Memoir of Mrs Hemans. By her SISTER. With a Portrait, fcap. 8vo, 5s.

HOLE. A Book about Roses: How to Grow and Show Them. By the Rev. Canon HOLE. With coloured Frontispiece by the Hon. Mrs Francklin. Seventh Edition, revised. Crown 8vo, 7s. 6d.

HOMER. The Odyssey. Translated into English Verse in the Spenserian Stanza. By PHILIP STANHOPE WORSLEY. Third Edition, 2 vols., fcap., 12s.

——— The Iliad. Translated by P. S. WORSLEY and Professor CONINGTON. 2 vols. crown 8vo, 21s.

HOME PRAYERS. By Ministers of the Church of Scotland and
Members of the Church Service Society. Fcap. 8vo, price 3s.

HOSACK. Mary Queen of Scots and Her Accusers. Containing a
Variety of Documents never before published. By JOHN HOSACK, Barrister-
at-Law. A New and Enlarged Edition, with a Photograph from the Bust on
the Tomb in Westminster Abbey. 2 vols. 8vo, £1, 1s.

INDEX GEOGRAPHICUS : Being a List, alphabetically arranged,
of the Principal Places on the Globe, with the Countries and Subdivisions of
the Countries in which they are situated, and their Latitudes and Longitudes.
Applicable to all Modern Atlases and Maps. Imperial 8vo, pp. 676, 21s.

JAMIESON. The Laird's Secret. By J. H. JAMIESON. In 2 vols.,
crown 8vo.

JEAN JAMBON. Our Trip to Blunderland ; or, Grand Excursion
to Blundertown and Back. By JEAN JAMBON. With Sixty Illustrations
designed by CHARLES DOYLE, engraved by DALZIEL. Fourth Thousand.
Handsomely bound in cloth, gilt edges, 6s. 6d. Cheap Edition, cloth, 3s. 6d.
In boards, 2s. 6d.

JOHNSON. The Scots Musical Museum. Consisting of upwards
of Six Hundred Songs, with proper Basses for the Pianoforte. Originally pub-
lished by JAMES JOHNSON ; and now accompanied with Copious Notes and
Illustrations of the Lyric Poetry and Music of Scotland, by the late WILLIAM
STENHOUSE; with additional Notes and Illustrations, by DAVID LAING and
C. K. SHARPE. 4 vols. 8vo, Roxburghe binding, £2, 12s. 6d.

JOHNSTON. The Chemistry of Common Life. By Professor
J. F. W. JOHNSTON. New Edition, Revised, and brought down to date. By
ARTHUR HERBERT CHURCH, M.A. Oxon. ; Author of 'Food: its Sources,
Constituents, and Uses ;' 'The Laboratory Guide for Agricultural Students ;'
'Plain Words about Water,' &c. Illustrated with Maps and 102 Engravings
on Wood. Complete in One Volume, crown 8vo, pp. 618, 7s. 6d.

—— Professor Johnston's Elements of Agricultural Chemistry
and Geology. Twelfth Edition, Revised, and brought down to date. By
CHARLES A. CAMERON, M.D., F.R.C.S.I., &c. Fcap. 8vo. 6s. 6d. .

—— Catechism of Agricultural Chemistry and Geology. An
entirely New Edition, revised and enlarged, by CHARLES A. CAMERON, M.D.,
F.R.C.S.I., &c. Eighty-first Thousand, with numerous Illustrations, 1s.

JOHNSTON. Patrick Hamilton : a Tragedy of the Reformation
in Scotland, 1528. By J. P. JOHNSTON. Crown 8vo, with Two Etchings by
the Author, 5s.

KEITH ELPHINSTONE. Memoir of the Honourable George
Keith Elphinstone, K.B., Viscount Keith of Stonehaven Marischal, Admiral
of the Red.—See ALEXANDER ALLARDYCE, at page 4.

KING. The Metamorphoses of Ovid. Translated in English Blank
Verse. By HENRY KING, M.A., Fellow of Wadham. College, Oxford, and of
the Inner Temple, Barrister-at-Law. Crown 8vo, 10s. 6d.

KINGLAKE. History of the Invasion of the Crimea. By A. W.
KINGLAKE. Cabinet Edition. Six Volumes, crown 8vo, at 6s. each. The Volumes respectively contain :—
    I. THE ORIGIN OF THE WAR between the Czar and the Sultan.
    II. RUSSIA MET AND INVADED. With 4 Maps and Plans.
    III. THE BATTLE OF THE ALMA. With 14 Maps and Plans.
    IV. SEBASTOPOL AT BAY. With 10 Maps and Plans.
    V. THE BATTLE OF BALACLAVA. With 10 Maps and Plans.
    VI. THE BATTLE OF INKERMAN. With 11 Maps and Plans.
    VII. WINTER TROUBLES. With Map.   [In the Press.

—— History of the Invasion of the Crimea. Vol. VI. Winter
Troubles. Demy 8vo, with a Map, 16s.

—— Eothen. A New Edition, uniform with the Cabinet Edition
of the 'History of the Crimean War,' price 6s.

KNOLLYS. The Elements of Field-Artillery. Designed for the Use of Infantry and Cavalry Officers. By HENRY KNOLLYS, Captain Royal Artillery; Author of 'From Sedan to Saarbrück,' Editor of 'Incidents in the Sepoy War,' &c. With Engravings. Crown 8vo, 7s. 6d.

LAKEMAN. What I saw in Kaffir-land. By Sir STEPHEN LAKEMAN (MAZHAR PACHA). Post 8vo, 8s. 6d.

LAVERGNE. The Rural Economy of England, Scotland, and Ireland. By LEONCE DE LAVERGNE. Translated from the French. With Notes by a Scottish Farmer. 8vo, 12s.

LEE. Lectures on the History of the Church of Scotland, from the Reformation to the Revolution Settlement. By the late Very Rev. JOHN LEE, D.D., LL.D., Principal of the University of Edinburgh. With Notes and Appendices from the Author's Papers. Edited by the Rev. WILLIAM LEE, D.D. 2 vols. 8vo, 21s.

LEE-HAMILTON. Poems and Transcripts. By EUGENE LEE-HAMILTON. Crown 8vo, 6s.

LEWES. The Physiology of Common Life. By GEORGE H. LEWES, Author of 'Sea-side Studies,' &c. Illustrated with numerous Engravings. 2 vols., 12s.

LOCKHART. Doubles and Quits. By Laurence W. M. LOCKHART. With Twelve Illustrations. Third Edition. Crown 8vo, 6s.

———— Fair to See : a Novel. Seventh Edition, crown 8vo, 6s.

———— Mine is Thine : a Novel. Seventh Edition, crown 8vo, 6s.

LORIMER. The Institutes of Law : A Treatise of the Principles of Jurisprudence as determined by Nature. By JAMES LORIMER, Regius Professor of Public Law and of the Law of Nature and Nations in the University of Edinburgh. New Edition, revised throughout, and much enlarged. 8vo, 18s.

———— The Institutes of the Law of Nations. A Treatise of the Jural Relation of Separate Political Communities. 8vo. [In the Press.

LYON. History of the Rise and Progress of Freemasonry in Scotland. By DAVID MURRAY LYON, Secretary to the Grand Lodge of Scotland. In small quarto. Illustrated with numerous Portraits of Eminent Members of the Craft, and Facsimiles of Ancient Charters and other Curious Documents. £1, 11s. 6d.

M'COMBIE. Cattle and Cattle-Breeders. By WILLIAM M'COMBIE, Tillyfour. A New and Cheaper Edition, 2s. 6d., cloth.

MACRAE. A Handbook of Deer-Stalking. By ALEXANDER MACRAE, late Forester to Lord Henry Bentinck. With Introduction by HORATIO ROSS, Esq. Fcap. 8vo, with two Photographs from Life. 3s. 6d.

M'CRIE. Works of the Rev. Thomas M'Crie, D.D. Uniform Edition. Four vols. crown 8vo, 24s.

———— Life of John Knox. Containing Illustrations of the History of the Reformation in Scotland. Crown 8vo, 6s. Another Edition, 3s. 6d.

———— Life of Andrew Melville. Containing Illustrations of the Ecclesiastical and Literary History of Scotland in the Sixteenth and Seventeenth Centuries. Crown 8vo, 6s.

———— History of the Progress and Suppression of the Reformation in Italy in the Sixteenth Century. Crown 8vo, 4s.

———— History of the Progress and Suppression of the Reformation in Spain in the Sixteenth Century. Crown 8vo, 3s. 6d.

———— Sermons, and Review of the 'Tales of My Landlord.' Crown 8vo, 6s.

———— Lectures on the Book of Esther. Fcap. 8vo, 5s.

M'INTOSH. The Book of the Garden. By CHARLES M'INTOSH, formerly Curator of the Royal Gardens of his Majesty the King of the Belgians, and lately of those of his Grace the Duke of Buccleuch, K.G., at Dalkeith Palace. Two large vols. royal 8vo, embellished with 1350 Engravings. £4, 7s. 6d.
Vol. I. On the Formation of Gardens and Construction of Garden Edifices. 776 pages, and 1073 Engravings, £2, 10s.
Vol. II. Practical Gardening. 868 pages, and 279 Engravings, £1, 17s. 6d.

MACKAY. A Manual of Modern Geography; Mathematical, Physical, and Political. By the Rev. ALEXANDER MACKAY, LL.D., F.R.G.S. New and Greatly Improved Edition. Crown 8vo, pp. 688. 7s. 6d.

―――― Elements of Modern Geography. 47th Thousand, revised to the present time. Crown 8vo, pp. 300, 3s.

―――― The Intermediate Geography. Intended as an Intermediate Book between the Author's 'Outlines of Geography,' and 'Elements of Geography.' Eighth Edition, crown 8vo, pp. 224, 2s.

―――― Outlines of Modern Geography. 142d Thousand, revised to the Present Time. 18mo, pp. 112, 1s.

―――― First Steps in Geography. 82d Thousand. 18mo, pp. 56. Sewed, 4d.; cloth, 6d.

―――― Elements of Physiography and Physical Geography. With Express Reference to the Instructions recently issued by the Science and Art Department. 19th Thousand. Crown 8vo, 1s. 6d.

―――― Facts and Dates; or, the Leading Events in Sacred and Profane History, and the Principal Facts in the various Physical Sciences. The Memory being aided throughout by a Simple and Natural Method. For Schools and Private Reference. New Edition, thoroughly Revised. Crown 8vo, 3s. 6d.

MACKENZIE. Studies in Roman Law. With Comparative Views of the Laws of France, England, and Scotland. By LORD MACKENZIE, one of the Judges of the Court of Session in Scotland. Fifth Edition, Edited by JOHN KIRKPATRICK, Esq., M.A. Cantab.; Dr Jur. Heidelb.; LL.B., Edin.; Advocate. 8vo. 12s.

MANNERS. Notes of an Irish Tour in 1846. By Lord JOHN MANNERS, M.P., G.C.B. New Edition, crown 8vo. 2s. 6d.

MARMORNE. The Story is told by ADOLPHUS SEGRAVE, the youngest of three Brothers. Third Edition. Crown 8vo, 6s.

MARSHALL. French Home Life. By FREDERIC MARSHALL.
CONTENTS: Servants.—Children.—Furniture.—Food.—Manners.—Language.—Dress.—Marriage. Second Edition. 5s.

MARSHMAN. History of India. From the Earliest Period to the Close of the India Company's Government; with an Epitome of Subsequent Events. By JOHN CLARK MARSHMAN, C.S.I. Abridged from the Author's larger work. Second Edition, revised. Crown 8vo, with Map, 6s. 6d.

MARTIN. Goethe's Faust. Translated by Sir THEODORE MARTIN, K.C.B. Second Edition, crown 8vo, 6s. Cheap Edition, 3s. 6d.

―――― The Works of Horace. Translated into English Verse, with Life and Notes. In 2 vols. crown 8vo, printed on hand-made paper, 21s.

―――― Poems and Ballads of Heinrich Heine. Done into English Verse. Second Edition. Printed on papier vergé, crown 8vo, 8s.

―――― Catullus. With Life and Notes. Second Edition, post 8vo, 7s. 6d.

―――― The Vita Nuova of Dante. With an Introduction and Notes. Second Edition, crown 8vo, 5s.

―――― Aladdin: A Dramatic Poem. By ADAM OEHLENSCHLAEGER. Fcap. 8vo, 5s.

MARTIN. Correggio: A Tragedy. By OEHLENSCHLAEGER. With
Notes. Fcap. 8vo, 3s
—— King Rene's Daughter : A Danish Lyrical Drama. By
HENRIK HERTZ. Second Edition, fcap., 2s. 6d.

MEIKLEJOHN. An Old Educational Reformer—Dr Bell. By
J. M. D. MEIKLEJOHN, M.A., Professor of the Theory, History, and Practice
of Education in the University of St Andrews. Crown 8vo, 3s. 6d.

MICHEL. A Critical Inquiry into the Scottish Language. With
the view of Illustrating the Rise and Progress of Civilisation in Scotland. By
FRANCISQUE-MICHEL, F.S.A. Lond. and Scot., Correspondant de l'Institut de
France, &c. In One handsome Quarto Volume, printed on hand-made paper,
and appropriately bound in Roxburghe style. Price 66s. *The Edition is
strictly limited to 500 copies, which will be numbered and allotted in the order of
application.*

MICHIE. The Larch : Being a Practical Treatise on its Culture
and General Management. By CHRISTOPHER YOUNG MICHIE, Forester, Cullen
House. Crown 8vo, with Illustrations. 7s. 6d.

MINTO. A Manual of English Prose Literature, Biographical
and Critical : designed mainly to show Characteristics of Style. By W. MINTO,
M.A., Professor of Logic in the University of Aberdeen. Second Edition,
revised. Crown 8vo, 7s. 6d.

—— Characteristics of English Poets, from Chaucer to Shirley.
Crown 8vo, 9s.

MITCHELL. Biographies of Eminent Soldiers of the last Four
Centuries. By Major-General JOHN MITCHELL, Author of 'Life of Wallenstein.'
With a Memoir of the Author. 8vo, 9s.

MOIR. Poetical Works of D. M. MOIR (Delta). With Memoir by
THOMAS AIRD, and Portrait. Second Edition, 2 vols. fcap. 8vo, 12s.

—— Domestic Verses. New Edition, fcap. 8vo, cloth gilt, 4s. 6d.

—— Lectures on the Poetical Literature of the Past Half-Cen-
tury. Third Edition, fcap. 8vo, 5s.

—— Life of Mansie Wauch, Tailor in Dalkeith. With 8
Illustrations on Steel, by the late GEORGE CRUIKSHANK. Crown 8vo. 3s. 6d.
Another Edition, fcap. 8vo, 1s. 6d.

MOMERIE. The Origin of Evil ; and other Sermons. Preached
in St Peter's, Cranley Gardens. By the Rev. A. W. MOMERIE, M.A., D.Sc.,
Fellow of St John's College, Cambridge ; Professor of Logic and Metaphysics
in King's College, London. Second Edition, enlarged. Crown 8vo, 5s.

—— Personality. The Beginning and End of Metaphysics, and
the Necessary Assumption in all Positive Philosophy. Crown 8vo, 3s.

MONTAGUE. Campaigning in South Africa. Reminiscences of
an Officer in 1879. By Captain W. E. MONTAGUE, 94th Regiment, Author of
'Claude Meadowleigh,' &c. 8vo, 10s. 6d.

MONTALEMBERT. Count de Montalembert's History of the
Monks of the West. From St Benedict to St Bernard. Translated by Mrs
OLIPHANT. 7 vols. 8vo, £3, 17s. 6d.

—— Memoir of Count de Montalembert. A Chapter of Recent French History. By Mrs OLIPHANT, Author of the 'Life of Edward
Irving,' &c. 2 vols. crown 8vo. £1, 4s.

MORE THAN KIN. By M. P. One volume, crown 8vo, 7s. 6d.

MURDOCH. Manual of the Law of Insolvency and Bankruptcy :
Comprehending a Summary of the Law of Insolvency, Notour Bankruptcy,
Composition - contracts, Trust-deeds, Cessios, and Sequestrations ; and the
Winding-up of Joint-Stock Companies in Scotland ; with Annotations on the
various Insolvency and Bankruptcy Statutes ; and with Forms of Procedure
applicable to these Subjects. By JAMES MURDOCH, Member of the Faculty of
Procurators in Glasgow. Fourth Edition, Revised and Enlarged, 8vo, £1.

NEAVES. A Glance at some of the Principles of Comparative Philology. As illustrated in the Latin and Anglican Forms of Speech. By the Hon. Lord NEAVES. Crown 8vo, 1s. 6d.

—— Songs and Verses, Social and Scientific. By an Old Contributor to 'Maga.' Fifth Edition, fcap. 8vo, 4s.

—— The Greek Anthology. Being Vol. XX. of 'Ancient Classics for English Readers.' Crown 8vo, 2s. 6d.

NEW VIRGINIANS, THE. By the Author of 'Estelle Russell,' 'Junia,' &c. In 2 vols., post 8vo, 18s.

NICHOLSON. A Manual of Zoology, for the Use of Students. With a General Introduction on the Principles of Zoology. By HENRY ALLEYNE NICHOLSON, M.D., F.R.S.E., F.G.S., &c., Professor of Natural History in the University of Aberdeen. Sixth Edition, revised and enlarged. Crown 8vo, pp. 866, with 452 Engravings on Wood, 14s.

—— Text-Book of Zoology, for the Use of Schools. Third Edition, enlarged. Crown 8vo, with 225 Engravings on Wood, 6s.

—— Introductory Text-Book of Zoology, for the Use of Junior Classes. Third Edition, revised and enlarged, with 136 Engravings, 3s.

—— Outlines of Natural History, for Beginners; being Descriptions of a Progressive Series of Zoological Types. Second Edition, with Engravings, 1s. 6d.

—— A Manual of Palæontology, for the Use of Students. With a General Introduction on the Principles of Palæontology. Second Edition. Revised and greatly enlarged. 2 vols. 8vo, with 722 Engravings, £2, 2s.

—— The Ancient Life-History of the Earth. An Outline of the Principles and Leading Facts of Palæontological Science. Crown 8vo, with numerous Engravings, 10s. 6d.

—— On the "Tabulate Corals" of the Palæozoic Period, with Critical Descriptions of Illustrative Species. Illustrated with 15 Lithograph Plates and numerous Engravings. Super-royal 8vo, 21s.

—— On the Structure and Affinities of the Genus Monticulipora and its Sub-Genera, with Critical Descriptions of Illustrative Species. Illustrated with numerous Engravings on wood and lithographed Plates. Super-royal 8vo, 18s.

—— Synopsis of the Classification of the Animal Kingdom. In one volume 8vo, with numerous Illustrations.

NICHOLSON. Communion with Heaven, and other Sermons. By the late MAXWELL NICHOLSON, D.D., Minister of St Stephen's, Edinburgh. Crown 8vo, 5s. 6d.

—— Rest in Jesus. Sixth Edition. Fcap. 8vo, 4s. 6d.

OLIPHANT. The Land of Gilead. With Excursions in the Lebanon. By LAURENCE OLIPHANT, Author of 'Lord Elgin's Mission to China and Japan,' &c. With Illustrations and Maps. Demy 8vo, 21s.

—— The Land of Khemi. Post 8vo, with Illustrations, 10s. 6d.

—— Traits and Travesties; Social and Political. Post 8vo, 10s. 6d.

—— Piccadilly: A Fragment of Contemporary Biography. With Eight Illustrations by Richard Doyle. Fifth Edition, 4s. 6d. Cheap Edition, in paper cover, 2s. 6d.

OLIPHANT. Historical Sketches of the Reign of George Second. By Mrs OLIPHANT. Third Edition, 6s.

—— The Story of Valentine; and his Brother. 5s., cloth.

—— Katie Stewart. 2s. 6d.

18 LIST OF BOOKS PUBLISHED BY

OLIPHANT. Salem Chapel. 2s. 6d., cloth.
—— The Perpetual Curate. 2s. 6d., cloth.
—— Miss Marjoribanks. 2s. 6d., cloth.
—— The Rector, and the Doctor's Family. 1s. 6d., cloth.
—— John : A Love Story. 2s. 6d., cloth.
OSBORN. Narratives of Voyage and Adventure. By Admiral
SHERARD OSBORN, C.B. 3 vols. crown 8vo, 12s.
OSSIAN. The Poems of Ossian in the Original Gaelic. With a
Literal Translation into English, and a Dissertation on the Authenticity of the
Poems. By the Rev. ARCHIBALD CLERK. 2 vols. imperial 8vo, £1, 11s. 6d.
OSWALD. By Fell and Fjord ; or, Scenes and Studies in Iceland.
By E. J. OSWALD. One Volume, post 8vo, with Illustrations. [In the Press.
PAGE. Introductory Text-Book of Geology. By DAVID PAGE,
LL.D., Professor of Geology in the Durham University of Physical Science,
Newcastle. With Engravings on Wood and Glossarial Index. Eleventh
Edition, 2s. 6d.
—— Advanced Text-Book of Geology, Descriptive and Indus-
trial. With Engravings, and Glossary of Scientific Terms. Sixth Edition, re-
vised and enlarged, 7s. 6d.
—— Geology for General Readers. A Series of Popular Sketches
in Geology and Palæontology. Third Edition, enlarged, 6s.
—— Chips and Chapters. A Book for Amateurs and Young
Geologists. 5s.
—— The Past and Present Life of the Globe. With numerous
Illustrations. Crown 8vo, 6s.
—— The Crust of the Earth : A Handy Outline of Geology.
Sixth Edition, 1s.
—— Economic Geology ; or, Geology in its relation to the Arts
and Manufactures. With Engravings, and Coloured Map of the British Islands.
Crown 8vo, 7s. 6d.
—— Introductory Text-Book of Physical Geography. With
Sketch-Maps and Illustrations. Edited by CHARLES LAPWORTH, F.G.S., &c.,
Professor of Geology and Mineralogy in the Mason Science College, Birming-
ham. 10th Edition. 2s. 6d.
—— Advanced Text-Book of Physical Geography. Second Edi-
tion. With Engravings. 5s.
PAGET. Paradoxes and Puzzles: Historical, Judicial, and Literary.
Now for the first time published in Collected Form. By JOHN PAGET, Barris-
ter-at-Law. 8vo, 12s.
PATON. Spindrift. By Sir J. NOEL PATON. Fcap., cloth, 5s.
—— Poems by a Painter. Fcap., cloth, 5s.
PATTERSON. Essays in History and Art. By R. H. PATTERSON.
8vo, 12s.
PAUL. History of the Royal Company of Archers, the Queen's
Body-Guard for Scotland. By JAMES BALFOUR PAUL, Advocate of the Scottish
Bar. Crown 4to, with Portraits and other Illustrations. £2, 2s.
PAUL. Analysis and Critical Interpretation of the Hebrew Text of
the Book of Genesis. Preceded by a Hebrew Grammar, and Dissertations on
the Genuineness of the Pentateuch, and on the Structure of the Hebrew Lan-
guage. By the Rev. WILLIAM PAUL, A.M. 8vo, 18s.
PETTIGREW. The Handy Book of Bees, and their Profitable
Management. By A. PETTIGREW. Fourth Edition, Enlarged, with Engravings. Crown 8vo, 3s. 6d.

PHILLIMORE. Uncle Z. By GREVILLE PHILLIMORE, Rector of Henley-on-Thames. Crown 8vo, 7s. 6d.

PHILOSOPHICAL CLASSICS FOR ENGLISH READERS.
Companion Series to Ancient and Foreign Classics for English Readers. Edited by WILLIAM KNIGHT, LL.D., Professor of Moral Philosophy, University of St Andrews. In crown 8vo volumes, with portraits, price 3s. 6d.

1. DESCARTES. By Professor Mahaffy, Dublin.
2. BUTLER. By the Rev. W. Lucas Collins, M.A., Honorary Canon of Peterborough.
3. BERKELEY. By Professor A. Campbell Fraser, Edinburgh.
4. FICHTE. By Professor Adamson, Owens College, Manchester.
5. KANT. By William Wallace, M.A., LL.D., Merton College, Oxford.

POLLOK. The Course of Time : A Poem. By ROBERT POLLOK, A.M. Small fcap. 8vo, cloth gilt, 2s. 6d. The Cottage Edition, 32mo, sewed, 8d. The Same, cloth, gilt edges, 1s. 6d. Another Edition, with Illustrations by Birket Foster and others, fcap., gilt cloth, 3s. 6d., or with edges gilt, 4s.

PORT ROYAL LOGIC. Translated from the French : with Introduction, Notes, and Appendix. By THOMAS SPENCER BAYNES, LL.D., Professor in the University of St Andrews. Eighth Edition, 12mo, 4s.

POST-MORTEM. Third Edition, 1s.

BY THE SAME AUTHOR.

The Autobiography of Thomas Allen. 3 vols. post 8vo, 25s. 6d.

POTTS AND DARNELL. Aditus Faciliores: An easy Latin Construing Book, with Complete Vocabulary. By A. W. POTTS, M.A., LL.D., Head-Master of the Fettes College, Edinburgh, and sometime Fellow of St John's College, Cambridge; and the Rev. C. DARNELL, M.A., Head-Master of Cargilfield Preparatory School, Edinburgh, and late Scholar of Pembroke and Downing Colleges, Cambridge. Seventh Edition, fcap. 8vo, 3s. 6d.

—— Aditus Faciliores Graeci. An easy Greek Construing Book, with Complete Vocabulary. Third Edition, fcap. 8vo, 3s.

PRINGLE. The Live-Stock of the Farm. By ROBERT O. PRINGLE. Third Edition, crown 8vo. [In the press.

PRIVATE SECRETARY. 3 vols. post 8vo, 25s. 6d.

PUBLIC GENERAL STATUTES AFFECTING SCOTLAND, from 1707 to 1847, with Chronological Table and Index. 3 vols. large 8vo, £3, 3s.

PUBLIC GENERAL STATUTES AFFECTING SCOTLAND, COLLECTION OF. Published Annually with General Index.

RAMSAY. Rough Recollections of Military Service and Society. By Lieut.-Col. BALCARRES D. WARDLAW RAMSAY. Two vols. post 8vo.

RANKINE. A Treatise on the Rights and Burdens Incident to the Ownership of Lands and other Heritages in Scotland. By JOHN RANKINE, M.A., Advocate. Large 8vo, 40s.

REID. A Handy Manual of German Literature. By M. F. REID. For Schools, Civil Service Competitions, and University Local Examinations. Fcap. 8vo, 3s.

REVOLT OF MAN. Post 8vo, 7s. 6d.

ROBERTSON. Orellana, and other Poems. By J. LOGIE ROBERTSON. Fcap. 8vo. Printed on hand-made paper. 6s.

—— Our Holiday Among the Hills. By JAMES and JANET LOGIE ROBERTSON. Fcap. 8vo, 3s. 6d.

RUSSELL. The Haigs of Bemersyde. A Family History. By JOHN RUSSELL. Large octavo, with Illustrations. 21s.

RUSTOW. The War for the Rhine Frontier, 1870 : Its Political
and Military History. By Col. W. Rustow. Translated from the German,
by John Layland Needham, Lieutenant R.M. Artillery. 3 vols. 8vo, with
Maps and Plans, £1, 11s. 6d.

SANDFORD and TOWNSEND. The Great Governing Families
of England. By J. Langton Sandford and Meredith Townsend. 2 vols.
8vo, 15s., in extra binding, with richly-gilt cover.

SCHETKY. Ninety Years of Work and Play. Sketches from the
Public and Private Career of John Christian Schetky, late Marine Painter in
Ordinary to the Queen. By his Daughter. Crown 8vo, 7s. 6d.

SCOTCH LOCH FISHING. By "Black Palmer." Crown 8vo,
interleaved with blank pages, 4s.

SCOTTISH NATURALIST, THE. A Quarterly Magazine of
Natural History. Edited by F. Buchanan White, M.D., F.L.S. Annual
Subscription, free by post, 4s.

SELLAR. Manual of the Education Acts for Scotland. By
Alexander Craig Sellar, Advocate. Seventh Edition, greatly enlarged,
and revised to the present time. 8vo, 15s.

SELLER and STEPHENS. Physiology at the Farm ; in Aid of
Rearing and Feeding the Live Stock. By William Seller, M.D., F.R.S.E.,
Fellow of the Royal College of Physicians, Edinburgh, formerly Lecturer on
Materia Medica and Dietetics ; and Henry Stephens, F.R.S.E., Author of ' The
Book of the Farm,' &c. Post 8vo, with Engravings, 16s.

SETON. Memoir of Alexander Seton, Earl of Dunfermline, Seventh
President of the Court of Session, and Lord Chancellor of Scotland. By
George Seton, M.A. Oxon.; Author of the 'Law and Practice of Heraldry in
Scotland,' &c. In 1 vol. 8vo.   [In the Press.

SHADWELL. The Life of Colin Campbell, Lord Clyde. Illustrated by Extracts from his Diary and Correspondence. By Lieutenant-
General Shadwell, C.B. 2 vols. 8vo. With Portrait, Maps, and Plans.
36s.

SIMPSON. Dogs of other Days : Nelson and Puck. By Eve
Blantyre Simpson. Fcap. 8vo, with Illustrations, 4s. 6d.

SMITH. The Pastor as Preacher ; or, Preaching in connection
with Work in the Parish and the Study ; being Lectures delivered at the
Universities of Edinburgh, Aberdeen, and Glasgow. By Henry Wallis
Smith, D.D., Minister of Kirknewton and East Calder ; one of the Lecturers
on Pastoral Theology appointed by the General Assembly of the Church of
Scotland. Crown 8vo, 5s.

SMITH. Italian Irrigation : A Report on the Agricultural Canals
of Piedmont and Lombardy, addressed to the Hon. the Directors of the East
India Company ; with an Appendix, containing a Sketch of the Irrigation System of Northern and Central India. By Lieut.-Col. R. Baird Smith, F.G.S.,
Captain, Bengal Engineers. Second Edition. 2 vols. 8vo, with Atlas in folio.
30s.

SMITH. Thorndale ; or, The Conflict of Opinions. By William
Smith, Author of 'A Discourse on Ethics,' &c. A New Edition. Crown
8vo, 10s. 6d.

—— Gravenhurst ; or, Thoughts on Good and Evil. Second
Edition, with Memoir of the Author. Crown 8vo, 8s.

—— A Discourse on Ethics of the School of Paley. 8vo, 4s.

—— Dramas. 1. Sir William Crichton. 2. Athelwold. 3.
Guidone. 24mo, boards, 3s.

SOUTHEY. Poetical Works of Caroline Bowles Southey. Fcap.
8vo, 5s.

—— The Birthday, and other Poems. Second Edition, 5s.

—— Chapters on Churchyards. Fcap., 2s. 6d.

SPEKE. What led to the Discovery of the Nile Source. By JOHN
HANNING SPEKE, Captain H.M. Indian Army. 8vo, with Maps, &c., 14s.

―――― Journal of the Discovery of the Source of the Nile. By
J. H. SPEKE, Captain H.M. Indian Army. With a Map of Eastern Equatorial
Africa by Captain SPEKE; numerous illustrations, chiefly from Drawings by
Captain GRANT; and Portraits, engraved on Steel, of Captains SPEKE and
GRANT. 8vo. 21s.

SPROTT. The Worship and Offices of the Church of Scotland ;
or, the Celebration of Public Worship, the Administration of the Sacraments,
and other Divine Offices, according to the Order of the Church of Scotland.
Being Lectures Delivered at the Universities of Aberdeen, Glasgow, St
Andrews, and Edinburgh. By GEORGE W. SPROTT, D.D., Minister of North
Berwick; one of the Lecturers on Pastoral Theology appointed by the
General Assembly of the Church of Scotland. Crown 8vo, 6s.

STARFORTH. Villa Residences and Farm Architecture : A Series
of Designs. By JOHN STARFORTH, Architect. 102 Engravings. Second Edition, medium 4to, £2, 17s. 6d.

STATISTICAL ACCOUNT OF SCOTLAND. Complete, with
Index, 15 vols. 8vo, £16, 16s.
Each County sold separately, with Title, Index, and Map, neatly bound in cloth,
forming a very valuable Manual to the Landowner, the Tenant, the Manufac-
turer, the Naturalist, the Tourist, &c.

STEPHENS. The Book of the Farm ; detailing the Labours of the
Farmer, Farm-Steward, Ploughman, Shepherd, Hedger, Farm-Labourer, Field-
Worker, and Cattleman. By HENRY STEPHENS, F.R.S.E. Illustrated with
Portraits of Animals painted from the life; and with 557 Engravings on Wood,
representing the principal Field Operations, Implements, and Animals treated
of in the Work. A New and Revised Edition, the third, in great part Re-
written. 2 vols. large 8vo, £2, 10s.

―――― The Book of Farm-Buildings ; their Arrangement and
Construction. By HENRY STEPHENS, F.R.S.E., Author of 'The Book of the
Farm ;' and ROBERT SCOTT BURN. Illustrated with 1045 Plates and En-
gravings. Large 8vo, uniform with 'The Book of the Farm,' &c. £1, 11s. 6d.

―――― The Book of Farm Implements and Machines. By J.
SLIGHT and R. SCOTT BURN, Engineers. Edited by HENRY STEPHENS. Large
8vo, uniform with 'The Book of the Farm,' £2, 2s.

―――― Catechism of Practical Agriculture. With Engravings. 1s.

STEWART. Advice to Purchasers of Horses. By JOHN STEWART,
V.S. Author of 'Stable Economy.' 2s. 6d.

―――― Stable Economy. A Treatise on the Management of
Horses in relation to Stabling, Grooming, Feeding, Watering, and Working.
Seventh Edition, fcap. 8vo, 6s. 6d.

STIRLING. Missing Proofs : a Pembrokeshire Tale. By M. C.
STIRLING, Author of 'The Grahams of Invermoy.' 2 vols. crown 8vo, 17s.

―――― The Minister's Son ; or, Home with Honours. 3 vols.
post 8vo. *[In the Press.*

STORMONTH. Etymological and Pronouncing Dictionary of the
English Language. Including a very Copious Selection of Scientific Terms.
For Use in Schools and Colleges, and as a Book of General Reference. By the
Rev. JAMES STORMONTH. The Pronunciation carefully Revised by the Rev.
P. H. PHELP, M.A. Cantab. Sixth Edition, with enlarged Supplement, con-
taining many words not to be found in any other Dictionary. Crown 8vo,
pp. 800. 7s. 6d.

―――― The School Etymological Dictionary and Word-Book.
Combining the advantages of an ordinary pronouncing School Dictionary and
an Etymological Spelling-book. Fcap. 8vo, pp. 254. 2s.

STORY. Graffiti D'Italia. By W. W. STORY, Author of 'Roba di
Roma.' Second Edition, fcap. 8vo, 7s. 6d.

STORY. Nero ; A Historical Play. Fcap. 8vo, 6s.

———— Vallombrosa. Post 8vo, 5s.

STRICKLAND. Lives of the Queens of Scotland, and English Princesses connected with the Regal Succession of Great Britain. By AGNES STRICKLAND. With Portraits and Historical Vignettes. 8 vols. post 8vo, £4, 4s.

STURGIS. John - a - Dreams. A Tale. By JULIAN STURGIS. New Edition, crown 8vo, 3s. 6d.

———— Little Comedies, Old and New. Crown 8vo, 7s. 6d.

———— Dick's Wandering. 3 vols., post 8vo, 25s. 6d.

SUTHERLAND. Handbook of Hardy Herbaceous and Alpine Flowers, for general Garden Decoration. Containing Descriptions, in Plain Language, of upwards of 1000 Species of Ornamental Hardy Perennial and Alpine Plants, adapted to all classes of Flower-Gardens, Rockwork, and Waters ; along with Concise and Plain Instructions for their Propagation and Culture. By WILLIAM SUTHERLAND, Gardener to the Earl of Minto ; formerly Manager of the Herbaceous Department at Kew. Crown 8vo, 7s. 6d.

TAYLOR. Destruction and Reconstruction : Personal Experiences of the Late War in the United States. By RICHARD TAYLOR, Lieutenant-General in the Confederate Army. 8vo, 10s. 6d.

TAYLOR. The Story of My Life. By the late Colonel MEADOWS TAYLOR, Author of 'The Confessions of a Thug,' &c. &c. Edited by his Daughter. New and cheaper Edition, being the Fourth. Crown 8vo, 6s.

THOLUCK. Hours of Christian Devotion. Translated from the German of A. Tholuck, D.D., Professor of Theology in the University of Halle. By the Rev. ROBERT MENZIES, D.D. With a Preface written for this Translation by the Author. Second Edition, crown 8vo, 7s. 6d.

THOMSON. Handy Book of the Flower-Garden : being Practical Directions for the Propagation, Culture, and Arrangement of Plants in Flower-Gardens all the year round. Embracing all classes of Gardens, from the largest to the smallest. With Engraved and Coloured Plans, illustrative of the various systems of Grouping in Beds and Borders. By DAVID THOMSON, Gardener to his Grace the Duke of Buccleuch, K.G., at Drumlanrig. Third Edition, crown 8vo, 7s. 6d.

———— The Handy Book of Fruit-Culture under Glass : being a series of Elaborate Practical Treatises on the Cultivation and Forcing of Pines, Vines, Peaches, Figs, Melons, Strawberries, and Cucumbers. With Engravings of Hothouses, &c., most suitable for the Cultivation and Forcing of these Fruits. Second Edition. Crown 8vo, with Engravings, 7s. 6d.

THOMSON. A Practical Treatise on the Cultivation of the Grape-Vine. By WILLIAM THOMSON, Tweed Vineyards. Ninth Edition, 8vo, 5s.

TOM CRINGLE'S LOG. A New Edition, with Illustrations. Crown 8vo, cloth gilt, 5s. Cheap Edition, 2s.

TRAILL. Recaptured Rhymes. Being a Batch of Political and other Fugitives arrested and brought to Book. By H. D. TRAILL. Crown 8vo, 5s.

TRANSACTIONS OF THE HIGHLAND AND AGRICUL-TURAL SOCIETY OF SCOTLAND. Published annually, price 5s.

TROLLOPE. The Fixed Period. By ANTHONY TROLLOPE. 2 vols., fcap. 8vo, 12s.

TULLOCH. Rational Theology and Christian Philosophy in England in the Seventeenth Century. By JOHN TULLOCH, D.D., Principal of St Mary's College in the University of St Andrews; and one of her Majesty's Chaplains in Ordinary in Scotland. Second Edition. 2 vols. 8vo, 28s.

TULLOCH. Some Facts of Religion and of Life. Sermons Preached before her Majesty the Queen in Scotland, 1866-76. Second Edition, crown 8vo, 7s. 6d.

—— The Christian Doctrine of Sin ; being the Croall Lecture for 1876. Crown 8vo, 6s.

—— Theism. The Witness of Reason and Nature to an All-Wise and Beneficent Creator. 8vo, 10s. 6d.

TYTLER. The Wonder-Seeker; or, The History of Charles Douglas. By M. FRASER TYTLER, Author of 'Tales of the Great and Brave,' &c. A New Edition. Fcap., 3s. 6d.

VIRGIL. The Æneid of Virgil. Translated in English Blank Verse by G. K. RICKARDS, M.A., and Lord RAVENSWORTH. 2 vols. fcap. 8vo, 10s.

WALFORD. Mr Smith : A Part of his Life. By L. B. WALFORD. Cheap Edition, 3s. 6d.

—— Pauline. Fifth Edition. Crown 8vo, 6s.

—— Cousins. Cheaper Edition. Crown 8vo, 6s.

—— Troublesome Daughters. Cheaper Edition. Crown 8vo, 6s.

—— Dick Netherby. Crown 8vo, 7s. 6d.

WARREN'S (SAMUEL) WORKS. People's Edition, 4 vols. crown 8vo, cloth, 18s. Or separately:—

Diary of a Late Physician. 3s. 6d. Illustrated, crown 8vo, 7s. 6d.

Ten Thousand A-Year. 5s.

Now and Then. The Lily and the Bee. Intellectual and Moral Development of the Present Age. 4s. 6d.

Essays : Critical, Imaginative, and Juridical. 5s.

WARREN. The Five Books of the Psalms. With Marginal Notes. By Rev. SAMUEL L. WARREN, Rector of Esher, Surrey ; late Fellow, Dean, and Divinity Lecturer, Wadham College, Oxford. Crown 8vo, 5s.

WELLINGTON. Wellington Prize Essays on "the System of Field Manœuvres best adapted for enabling our Troops to meet a Continental Army." Edited by General Sir EDWARD BRUCE HAMLEY, K.C.M.G. 8vo, 12s. 6d.

WESTMINSTER ASSEMBLY. Minutes of the Westminster Assembly, while engaged in preparing their Directory for Church Government, Confession of Faith, and Catechisms (November 1644 to March 1649). Printed from Transcripts of the Originals procured by the General Assembly of the Church of Scotland. Edited by the Rev. ALEX. T. MITCHELL, D.D., Professor of Ecclesiastical History in the University of St Andrews, and the Rev. JOHN STRUTHERS, LL.D., Minister of Prestonpans. With a Historical and Critical Introduction by Professor Mitchell. 8vo, 15s.

WHITE. The Eighteen Christian Centuries. By the Rev. JAMES WHITE, Author of 'The History of France.' Seventh Edition, post 8vo, with Index, 6s.

—— History of France, from the Earliest Times. Sixth Thousand, post 8vo, with Index, 6s.

WHITE. Archæological Sketches in Scotland—Kintyre and Knap-
dale. By Captain T. P. WHITE, R.E., of the Ordnance Survey. With numerous Illustrations. 2 vols. folio, £4, 4s. Vol. I., Kintyre, sold separately,
£2, 2s.

WILLS AND GREENE. Drawing-room Dramas for Children. By
W. G. WILLS and the Hon. Mrs GREENE. Crown 8vo, 6s.

WILSON. The "Ever-Victorious Army:" A History of the
Chinese Campaign under Lieut.-Col. C. G. Gordon, and of the Suppression of
the Tai-ping Rebellion. By ANDREW WILSON, F.A.S.L. 8vo, with Maps, 15s.

——— The Abode of Snow: Observations on a Journey from
Chinese Tibet to the Indian Caucasus, through the Upper Valleys of the
Himalaya. New Edition. Crown 8vo, with Map, 10s. 6d.

WILSON. Works of Professor Wilson. Edited by his Son-in-Law,
Professor FERRIER. 12 vols. crown 8vo, £2, 8s.

——— Christopher in his Sporting-Jacket. 2 vols., 8s.

——— Isle of Palms, City of the Plague, and other Poems. 4s.

——— Lights and Shadows of Scottish Life, and other Tales. 4s.

——— Essays, Critical and Imaginative. 4 vols., 16s.

——— The Noctes Ambrosianæ. Complete, 4 vols., 14s.

——— The Comedy of the Noctes Ambrosianæ. By CHRISTOPHER
NORTH. Edited by JOHN SKELTON, Advocate. With a Portrait of Professor
Wilson and of the Ettrick Shepherd, engraved on Steel. Crown 8vo, 7s. 6d.

——— Homer and his Translators, and the Greek Drama. Crown
8vo, 4s.

WINGATE. Annie Weir, and other Poems. By DAVID WINGATE.
Fcap. 8vo, 5s.

——— Lily Neil. A Poem. Crown 8vo, 4s. 6d.

WORSLEY. Poems and Translations. By PHILIP STANHOPE
WORSLEY, M.A. Edited by EDWARD WORSLEY. Second Edition, enlarged.
Fcap. 8vo, 6s.

WYLDE. A Dreamer. By KATHARINE WYLDE. In 3 vols.,
post 8vo, 25s. 6d.

YOUNG. Songs of Béranger done into English Verse. By WILLIAM
YOUNG. New Edition, revised. Fcap. 8vo, 4s. 6d.

YULE. Fortification: for the Use of Officers in the Army, and
Readers of Military History. By Col. YULE, Bengal Engineers. 8vo, with
numerous Illustrations, 10s. 6d.

www.ingramcontent.com/pod-product-compliance
Lightning Source LLC
Chambersburg PA
CBHW020551270326
41927CB00006B/794